SpringerBriefs in Applied Sciences and Technology

SpringerBriefs present concise summaries of cutting-edge research and practical applications across a wide spectrum of fields. Featuring compact volumes of 50 to 125 pages, the series covers a range of content from professional to academic.

Typical publications can be:

- A timely report of state-of-the art methods
- An introduction to or a manual for the application of mathematical or computer techniques
- A bridge between new research results, as published in journal articles
- A snapshot of a hot or emerging topic
- An in-depth case study
- A presentation of core concepts that students must understand in order to make independent contributions

SpringerBriefs are characterized by fast, global electronic dissemination, standard publishing contracts, standardized manuscript preparation and formatting guidelines, and expedited production schedules.

On the one hand, **SpringerBriefs in Applied Sciences and Technology** are devoted to the publication of fundamentals and applications within the different classical engineering disciplines as well as in interdisciplinary fields that recently emerged between these areas. On the other hand, as the boundary separating fundamental research and applied technology is more and more dissolving, this series is particularly open to trans-disciplinary topics between fundamental science and engineering.

Indexed by EI-Compendex, SCOPUS and Springerlink.

Azman Ismail · Fatin Nur Zulkipli ·
Husna Sarirah Husin · Andreas Öchsner

Editors

Tech Horizons

Unveiling Future Technologies

 Springer

Editors
Azman Ismail
Maritime Engineering Technology
and Centre for Women Advancement
and Leadership
Universiti Kuala Lumpur Malaysian
Institute of Marine Engineering Technology
Lumut, Perak, Malaysia

Husna Sarirah Husin
Taylor's University
Subang Jaya, Selangor, Malaysia

Fatin Nur Zulkipli
Information Science Studies
College of Computing, Informatics
and Mathematics
Universiti Teknologi MARA
Machang, Kelantan, Malaysia

Andreas Öchsner
Faculty of Mechanical Engineering
Esslingen University of Applied Science
Esslingen am Neckar, Baden-Württemberg,
Germany

ISSN 2191-530X   ISSN 2191-5318   (electronic)
SpringerBriefs in Applied Sciences and Technology
ISBN 978-3-031-63325-6   ISBN 978-3-031-63326-3   (eBook)
https://doi.org/10.1007/978-3-031-63326-3

Preface

This book assembles a varied array of chapters, each delving into a distinct aspect of innovation and its practical applications. Readers will explore cutting-edge technologies and applicable techniques that aimed at enhancing academic performance. "Tech Horizon" provides an enthralling exploration of the diverse and transformative vistas within the domain of modern technology.

Lumut, Malaysia	Azman Ismail
Machang, Malaysia	Fatin Nur Zulkipli
Subang Jaya, Malaysia	Husna Sarirah Husin
Esslingen am Neckar, Germany	Andreas Öchsner

Preface

Contents

The Design and Development of a 2D Platform Narrative Digital Game: What Happened to Aubrey

Norshahila Ibrahim, Nurul Anisa Yusoff, Farzana Khairuzzaman, Noor Hidayah Azmi, Erni Marlina Saari, Albert Yakobus Chandra, and Mimi Dalina Ibrahim

Abstract Schizophrenia is one of the mental health disorders that has gone unnoticed due to a lack of understanding about the illness. Furthermore, there is a lack of digital games that can assist in spreading the word about schizophrenia. Thus, the WHTA, a 2D platform narrative game, was produced to raise awareness about schizophrenia among Malaysian citizens. The player takes on the role of Charlie, an 18-year-old young adult on a quest to find his companion Aubrey, who has gone missing. The players need to escape each level by finding enough clues that will reveal information about schizophrenia by fighting obstacles. The GDLC had been used as a guideline in the development process which consists of pre-production, production, post-production, and testing. Then, the playtesting had been conducted using the

N. Ibrahim (✉) · N. A. Yusoff · F. Khairuzzaman · N. H. Azmi · E. M. Saari
Faculty of Computing and Meta-Technology, Universiti Pendidikan Sultan Idris, Tanjong Malim, Perak, Malaysia
e-mail: shahila@meta.upsi.edu.my

N. A. Yusoff
e-mail: nurulanisayusoff@gmail.com

F. Khairuzzaman
e-mail: farzanakhairuzzaman12@gmail.com

N. H. Azmi
e-mail: hidayah@meta.upsi.edu.my

E. M. Saari
e-mail: marlina@meta.upsi.edu.my

A. Y. Chandra
Universitas Mercu Buana Yogyakarta, Yogyakarta, Indonesia
e-mail: albert.ch@mercubuana-yogya.ac.id

M. D. Ibrahim
Kolej Poly-Tech Mara Batu Pahat, Sri Gading, Malaysia
e-mail: mimi_dalina@gapps.kptm.edu.my

A. Ismail et al. (eds.), *Tech Horizons*,
SpringerBriefs in Applied Sciences and Technology,
https://doi.org/10.1007/978-3-031-63326-3_1

1

UEQ that measures the six elements including (i) attractiveness, (ii) perspicuity, (iii) efficiency, (iv) dependability, (v) stimulation, and (vi) novelty. A total of 33 respondents were involved during the playtesting. The game received positive reviews and comments in terms of the game design. For future work, the game can be enhanced including the game content, game-level design, and game feedback.

Keywords 2D platform game · Narrative game · Schizophrenia · Mental health · Digital game

1 Introduction

People are becoming more aware of mental health issues and how they can negatively impact another's life. Schizophrenia continues to be a common mental health issue that is rarely discussed and is still the subject of many misconceptions. However, to completely comprehend this subject, it must be digested. The National Health and Morbidity Survey, conducted by the Ministry of Health (MOH), the prevalence of mental health problems among people aged 16 and up was 29.2%, or approximately 4.2 million people [1, 2]. According to this figure, one out of every three Malaysians has had mental health issues. The study also indicated that most respondents were unfamiliar with and did not actually comprehend the term mental health, particularly the meaning of mental illness.

2 Literature Review

2.1 Schizophrenia

Schizophrenia is a complex mental disorder characterized by a disintegration of thought processes, emotions, and behavior, leading to delusions, hallucinations, and impaired social functioning. It often emerges during early adulthood and has significant impacts on cognition, perception, and interpersonal relationships, requiring specialized medical and psychological treatment. According to Galderisi and Mucci [3], schizophrenia is considered a psychosis which is a type of mental illness where the individuals who suffer have difficulty differentiating between imagination and reality which sometimes results in a loss of touch with reality called psychotic episodes. Schizophrenia can alter an individual's thoughts, feelings, and behavior. The symptoms of schizophrenia are divided into two which are positive (altered perception, abnormal thinking, excessive or abnormal behavior, and disorganized speech) and negative symptoms (diminished or absent ability to function properly).

2.2 Narrative Game

A narrative game is known to be focused deeply on story structures. Narrative-driven games are a massive part of the industry, with the main aim of telling stories [4]. Some narrative-driven games receive higher ratings than others, showing that players have preferences to enjoy a game. Moreover, the aim is to identify the player preferences on elements that affect their perception of the narrative. Thus, it is quite popular to be used for games whose main objective is to deliver its story to the player. The debate about whether video games are a narrative medium may be due to their origins as they initially gained popularity as a form of entertainment in the late 1970s.

3 Methodology

The game development life cycle (GDLC) had been chosen as a guideline for the development process. Many previous works had been chosen for the GDLC in project management and guidelines [5–7]. GDLC consists of four phases which are (i) pre-production, (ii) production, (iii) post-production, and (iv) testing.

3.1 Pre-production

The three steps of pre-production include (i) ideation and brainstorming, (ii) developing story concepts, and (iii) creating gameplay and game design. The studies on current issues in mental health had been done by doing literature reviews in journals, conference proceedings, and news updates. A meeting had been set up with one of the subject matter experts in psychology from UPSI to discuss more on the identified issues. After that, the game storyline, games flow, and game storyboard had been developed. A game concept design (GCD) had been created consisting of the completion of the gameplay, encompassing the game's appearance and feel. Table 1 shows the summary of GCD.

3.2 Production

The development of characters, game assets, and animation for cut scenes had been created using Ibis Paint and Procreate software. Each animation can be saved as MP4 files or PNG sequence files and transferred into the computer to be compiled as a video. Finally, all animation, cutscenes, video, and sound effects will be edited in Adobe Premiere Pro for the final product. The Unity game engine was used to create the game after the production of the 2D characters and game assets was completed.

Table 1 Summary of game concept design (GCD)

Item	Description
Genre	2D platform narrative game
Issue/problem	Schizophrenia continues to be a common mental health issue that is rarely discussed and is still the subject of many misconceptions
Unique selling point	The game content is about giving awareness of schizophrenia
Goals	Find Aubrey and escape from all enemies
Rules	1. Avoid or defeat enemies 2. Find all the information at every level 3. Search for the keys to escape to the next level
Narrative	Explore the world with Charlie, who is desperately trying to find his missing friend, Aubrey, search notes for information, and escape the foreign world filled with (enemies) by answering most questions correctly
Challenges	Find the notes, get the correct answer, get the key to unlocking levels, and the number of enemies
Feedback	1. Information on schizophrenia will appear after the player collects the clues 2. The player will proceed to the next levels after answering all questions at the end of the levels 3. The player will be able to proceed to the unlocked door using the keys provided in level 2 4. Players' lives will be deducted if collide with the enemies

Game Prototype

The player plays the character named Charlie, who needs to search notes for information and escape the foreign world filled with enemies by answering most questions correctly. They need to search for the keys to escape to the next level. Figure 1 shows the screenshot of the WHTA. There are three questions in each level regarding schizophrenia that players need to answer before they can proceed to the next level. Each level has a different number of enemies to increase the difficulties and challenges as the player escapes from them. The player will collect the clues that have been scattered throughout the game, and the clues will expose the schizophrenia information.

3.3 Post-production

In the post-production phase, after the game had been completely developed, a functionality test had been conducted among researchers to ensure that each function in the game worked as planned.

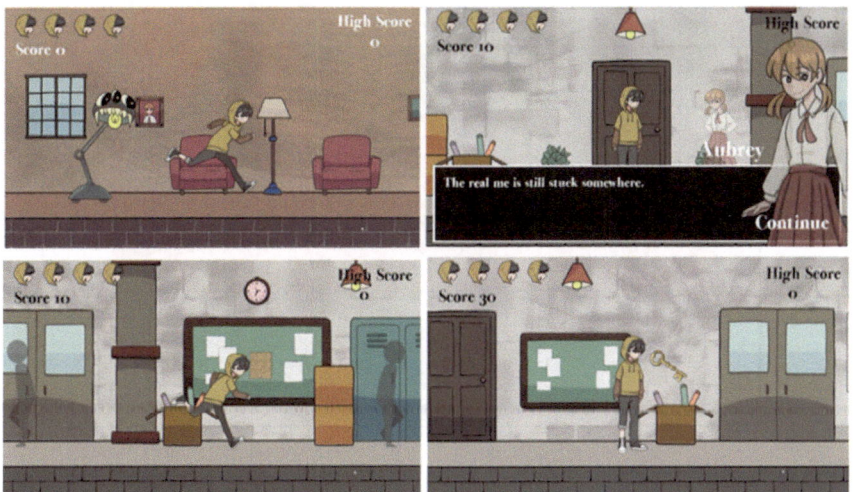

Fig. 1 Screenshots of the WHTA

3.4 Testing

The playtesting had been conducted to gain feedback on the user experience by recruiting the potential target players to try out the developed game. There were 33 respondents' feedbacks recorded through Google Forms. The respondents had more than 3 years of experience in playing games and knowledge regarding schizophrenia. 20 min had been given to the respondents to play and explore the game. At the end of the session, the respondents needed to scan a QR code to answer the questionnaire.

Instrument

The questionnaire consists of three sections which are (i) demographic profile, (ii) user experience questionnaire, and (iii) comments and suggestions. A user experiences questionnaire (UEQ) that had been used as an instrument to collect the data was adapted from [8]. Many previous works had been used UEQ as an instrument to measure the subjective impression of users toward the user experience of products [9, 10]. The adapted UEQ consists of six items asked including (i) attractiveness, (ii) perspicuity, (iii) efficiency, (iv) dependability, (v) stimulation, and (vi) novelty. The collected data had been analyzed using the template provided by the UEQ.

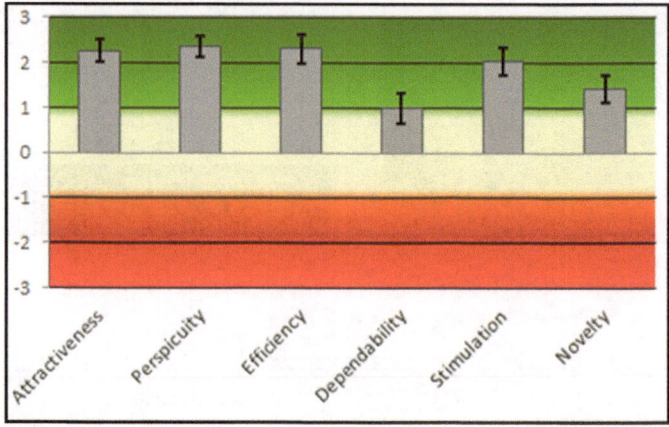

Fig. 2 Results of the UEQ for WHTA

4 Results and Discussion

The results show that the value for attractiveness is 2.267, perspicuity is 2.369, efficiency is 2.310, dependability is 1.000, stimulation is 2.032, and novelty is 1.429. Figure 2 shows the results of the UEQ of WHAT. According to Rauschenberger [11], the range scales between − 3 indicates the horribly bad response meanwhile + 3 indicates the extremely good. The value for all UEQ scales is in the range scale + 3; thus, the respondents give positive feedback for the WHTA.

Based on the playtester's comments and suggestions, the game received positive reviews and feedback. However, there are some parts of the game functionality that need to be fixed and improved, especially gameplay, character animation, and game theme. The playtester liked the overall design of the game, especially the character design. They also suggested adding more game-level designs in the future. The overall atmosphere and style are very much suitable for the theme of the game. But in order to understand schizophrenia thru playing games, creating more suitable mechanics need to be more precise.

5 Conclusion

For future works, the game can be improvised and enhanced in terms of game design and gameplay including game content, game-level design, and game feedback.

Acknowledgements The authors wish to thank the Faculty of Computing and Meta-Technology, Universiti Pendidikan Sultan Idris (UPSI) for the lab facilities to complete this project. The authors also want to thank everyone directly or indirectly involved in this project.

References

1. M.F. bin Hassan, N.M. Hassan, E.S. Kassim, M.I. Hamzah, Issues and challenges of mental health in Malaysia. Int. J. Acad. Res. Bus. Soc. Sci. **8**, 1685–1696 (2018)
2. A. Perveen, P. Govindasamy, E. Morgul, N.A.H. Abas, P. Kee, H. Hamzah, Life style behaviors as contributing factors of depression among university students. Int. Rev. Soc. Sci. **7**, 4 (2019)
3. S. Galderisi, A. Mucci, R.W. Buchanan, C. Arango, Negative symptoms of schizophrenia: new developments and unanswered research questions. Lancet Psychiatry **5**, 664–677 (2018)
4. S.N. Acarsoy, Effects of interactivity on narrative-driven games: a heuristic approach for narrative driven games (2021)
5. L.C. Xian, N. Ibrahim, N.H. Azmi, E.M. Saari, F.M. Razali, The development of an augmented reality game KANJI write for beginners. J. ICT Educ. **8**, 79–92 (2021). https://doi.org/10.37134/jictie.vol8.2.8.2021
6. S.N.F.N. Azhar, E.M. Saari, N. Ibrahim, F.M. Razali, N.H. Azmi: The engagement of visual novel digital game entitled single mingle: career versus marriage. J. ICT Educ. (JICTIE) **9**, 131–142 (2022). https://doi.org/10.37134/jictie.vol9.1.11.2022
7. N.S. Ahmad Anuar, M.S. Mahmud Fauzi, N.H. Azmi, N. Ibrahim, E.M. Saari, F. Mohd Razali, Design and development of periodic table game for students in secondary school. Int. J. Creative Multimedia **2**, 15–29 (2021). https://doi.org/10.33093/ijcm.2021.2.2.2
8. M. Schrepp, A. Hinderks, J. Thomaschewski, Applying the user experience questionnaire (UEQ) in different evaluation scenarios, in *Design, User Experience, and Usability. Theories, Methods, and Tools for Designing the User Experience*, ed. by A. Marcus (Springer International Publishing, 2014), pp. 383–392
9. W. Ortiz, D. Castillo, L. Wong, Mobile application: a serious game based in gamification for learning mathematics in high school students, in *2022 31st Conference of Open Innovations Association (FRUCT)* (2022)
10. N. Ibrahim, W.F.W. Ahmad, A. Shafie, User experience study on folktales mobile application for children's education, in *9th International Conference on Next Generation Mobile Applications, Services and Technologies (NGMAST)* (2015), pp. 353–358
11. M. Rauschenberger, M. Schrepp, M. Perez-Cota, S. Olschner, J. Thomaschewski, Efficient measurement of the user experience of interactive products. How to use the user experience questionnaire (UEQ). Example: Spanish language version. Int. J. Artif. Intelli. Interactive Multimedia **2**, 39–45 (2013). https://doi.org/10.9781/ijimai.2013.215

A Short Communication on the Application of MATLAB in Material Science

Md. Gulam Smdani, Muhammad Remanul Islam, Mohd Ismail Yusuf, Sairul Izwan Safie, Ahmad Naim Ahamad Yahaya, and Amin Firouzi

Abstract With the ongoing swift growth of computer technology and the growing degree of automation in manufacturing processes, it is apparent that software applications will continue to take over a rising number of jobs and roles. MATLAB is a widely used software tool in chemical processes to enhance product quality, reduce manufacturing costs, shorten operational time, and reduce environmental impact. This study investigated the application of MATLAB-based simulation in image processing algorithms for materials' characteristics. MATLAB is a useful tool for precisely characterizing material properties with less simulation time, accuracy, easy and quick visualization, and lower labor and expense.

Keywords MATLAB · Simulation · Image processing · Material characterization

Md. G. Smdani
South Dakota Mines, Rapid City, USA
e-mail: sumdani.cep06@yahoo.com

M. R. Islam (✉) · S. I. Safie
Plant Engineering Technology Section, Universiti Kuala Lumpur Malaysian Institute of Industrial Technology, Pasir Gudang, Johor Bahru, Malaysia
e-mail: muhammad.remanul@unikl.edu.my

S. I. Safie
e-mail: sairul@unikl.edu.my

M. I. Yusuf
Instrumentation and Control Engineering, Universiti Kuala Lumpur Malaysian Institute of Industrial Technology, Pasir Gudang, Johor Bahru, Malaysia
e-mail: ismail@unikl.edu.my

A. N. A. Yahaya
Institute of Postgraduate Studies, Universiti Kuala Lumpur, Kuala Lumpur, Malaysia
e-mail: ahmadnaim@unikl.edu.my

A. Firouzi
GE Appliances, Loushvelli, KY, USA
e-mail: firouzi.amin@gmail.com

© The Author(s), under exclusive license to Springer Nature Switzerland AG 2024
A. Ismail et al. (eds.), *Tech Horizons*,
SpringerBriefs in Applied Sciences and Technology,
https://doi.org/10.1007/978-3-031-63326-3_2

1 Introduction

Chemical engineering is a crucial foundational field in the engineering discipline. It is vital to the advancement of science and technology as well as the advancement of civilizations. Chemical studies, for example, play a crucial part in chemical engineering. Experimental studies have led to advancements in chemical engineering. Processing experimental data has a significant impact and is a crucial part of chemical research. Processing the experimental data in a proper and effective manner can provide the desired chemical products. Chemical research professionals, on the other hand, performed the analysis of data manually in the past. This method is inefficient in terms of labor and working hours, and it frequently contains errors. Chemical engineering experts must create novel data processing technologies in this environment. Chemical engineering has benefited immensely from the successful establishment of MATLAB. It has reduced and streamlined the calculation stages, as well as increased the accuracy of research data analysis and considerably improved operational efficiencies [1].

The benefit of software applications in chemical industries has already been acknowledged by today's fast-changing business contexts. Software offers a shared platform for activity coordination, effective client interaction, actual information exchange for multiple actions, and automated scheduled operations. These features reduce waste and delays, improve governance, increase flexibility, promote corporate integration, and lower operational costs.

2 MATLAB

The MathWorks, Inc. of Natick, MA produces MATLAB, which stands for MATrix LABoratory. MATLAB is a strong and high-performance programming language for scientific research. MATLAB/Simulink is a computer-based engineering and scientific toolkit that provides useful interactive presentations and the ability to quickly generate various simulations based on theory. Through a sophisticated graphical user interface (GUI), it is a unified scientific software platform that integrates numeric computation, complex graphics, and visualization [2]. It combines computation, visualization, and coding in a user-friendly interface with problems and solutions written in common mathematical terminology. Math and computing, algorithm creation, modeling, simulation, and design, data analysis, exploration, and visualization, scientific and technical graphics, and systems integration like as graphical user interface building are all examples of typical applications.

3 Application of MATLAB in Materials Science

3.1 *Image Processing Using MATLAB*

Recent technological advancements have sparked a surge in enthusiasm in image processing systems as a study topic. Image processing systems, in general, are fast growing in popularity among technologies, and they are an attractive research area in chemical engineering. This approach entails performing operations on a picture with the aim of improving its quality or obtaining valuable and meaningful information from it. It refers to a signal processing function on an image in which the image serves as an input signal with output features that are identical to the original/input image. Image processing can be defined as a mathematical operation performed by a software on a two-dimensional photo, as well as a collection of signal processing processes performed on an image to convert it to a binary code for the purpose of improving quality and extracting usable information [3]. MATLAB-based image processing can be employed to provide the needed quality appropriate for a two-dimensional image, even if numerous image processing equipment such as cameras, X-ray devices, ultrasonic devices, and electron microscopes are utilized to create digital images. MATLAB is an advanced-level programming language with attractive traits that are important to both the computational and engineering fields, as well as algorithms with broad application possibilities [4]. Many researchers now have abundant access to multimedia software and hardware devices executing image processing programs, including 3D visualization, thanks to the continuous advancement of technology.

3.2 *Mechanism of Image Processing with MATLAB*

Image processing is a method of interpreting an image through pixel modification. Pixels are arranged in rows and columns to form an image. Each color is made up of a mix of red, green, and blue values ranging from 0 to 255. When all three colors are 0, the final color is black. When all three colors have the maximum value, 255, the final color seems to be white [5]. An image is treated as a three-dimensional matrix in MATLAB. Rows and columns are the first two dimensions, whereas the third dimension has three values: red, blue, and green. The programming language of MATLAB was used for coding since it includes built-in functions for image processing [6]. It is sometimes required to use filters to reduce noise and distortion from an image before it can be interpreted. It is necessary to comprehend the various parts of image processing to process the images. Image enhancement, restoration, and segmentation are all stages in the image processing method. Image enhancement is used to reduce noise, de-blur out-of-focus images, and better the appearance of an image. To eliminate intermittent interference and pinpoint different items in the primary image, image restoration and segmentation are performed.

3.3   Surface Velocity Profile of Polymer

MATLAB®-based image processing algorithms can be used high-speed polymer melt extrusion film casting (EFC) approach. The necking fault and surface velocity profiles of the generated polymer films were successfully identified and analyzed. The degree of necking in an EFC method was assessed using an image processing technique. The strategy is based on the examination of a series of image frames captured with a standard CCD camera over a defined EFC process targeted region. Large-scale, high-quality photographs are cropped to the right dimensions. The Canny method with multiple morphological procedures is used to detect clean edges. The image series is then examined with a MATLAB®-based image processing toolkit, where a bespoke algorithm is created and run to identify the edges of the produced polymer film to evaluate the necking problem. This research concludes that image processing approaches are useful for assessing both the necking fault and the accompanying velocity profiles in melted extruded films.

3.4   Void Fraction Identification

Sukamta and Sudarja [7] observed the void pattern of flow of air-pure water and glycerin using a MATLAB-based image processing technique. The experiment was carried out on a tiny pipe having a diameter of 1.6 mm with a length of 130 mm that was mounted at a 45-degree angle to the horizontal. A composition of air-pure water and glycerin with ratios of 40, 50, 60, and 70% is employed as the fluid. The flow pattern was recorded with a Nikon J4 camera at 1200 frames per second, with gas velocity (JG) at rates of 0.025–66.3 m/s and fluid velocity (JL) at rates of 0.033–4.935 m/s. It can also be described that raising the glycerin content causes the bubble to grow, like a kind of plug flow regime, and reduces the frequency of occurrence. Using a MATLAB-based image processing approach, they were able to successfully understand the characteristics of void fraction.

3.5   Material Characterization

Computer-aided applications have grown in popularity in tandem with the advancement of computer technology. Before any materials are assembled on a production process, there is a need for speedy, comprehensive analysis systems, simulation, and visualization approaches. Computer-based software is effective and accurate to characterize materials in chemical processes. Physical properties, chemical properties, microstructure and phase distribution, and texture analysis are all examples of material characterization that may be done with computer software [8]. Software is used to identify physical properties such as surface area, pore size, pore size distribution, pore structure, and pore volume. Moreover, software is also used to monitor chemical properties such as elemental analysis, thermal gravimetric analysis, crystallite size of catalytic species, and catalyst surface composition.

3.6　Particle Size Analysis

One of the most fundamental physical features of a particulate material is particle size distribution. The observed particle diameter value is determined by the particle shape and the particle size technology applied. Laser-based techniques, sieve analysis, X-ray technologies, and scanning electron microscope (SEM) are some of the methods used to determine particle size. Particle size and shape, on the other hand, are not always consistent among samples. This morphological variation can affect the resulting average particle dimension value, limiting tests and applications. Ortega et al. demonstrated the relevance of using the correct particle size for estimating the mass flow rate outflow of particles from a cavity-type falling particle receiver in recent research. Ortega et al. created a MATLAB analysis program that automatically processes photos based on the required camera and lens attributes embedded inside the same image file, necessitating the user to merely input the camera mounting position [9]. When compared to data from ImageJ imaging tools and a sieve analysis, the results showed a superior agreement. The MATLAB-based script technique delivered (i) morphological resilience for the analysis, (ii) high precision and rapid tests, and (iii) a cost-effective solution.

3.7　Surface Morphology

Surface metrology may provide a useful character of materials. In some ways, researchers were intended at defining the elements of a "functional surface metrology", in which the elements of surface topography could be evaluated based on the industrial processes that caused their appearance, or to provide quantifiable data on light scattering and color and texture of material surfaces [10]. The use of morphological features and synthesis methods for 3D-processing and identification of material surfaces was highlighted in some previous research works. Images, signals, and N-dimensional maps are all explored using nonlinear geometrical methods. Among many other image processing approaches, mathematical morphology is special. Mathematical morphology is employed in computational materials science to characterize the atomic cluster configuration in Monte Carlo/molecular dynamics simulations involving millions of atoms [11]. Image processing tools like ImageJ have a basic set of elementary morphological functions (erosion, dilation, opening, closure) as well as advanced techniques (e.g., watershed transform) [12].

4 Conclusion

Chemical processes must be designed with the supply of good quality products, low energy demands, and minimal environmental effect. Many scientists have focused on optimization methodologies to simulate chemical processes with the lowest cost and energy demands. Regardless of the optimization method employed, a modeling of the specific chemical process is required. Various studies investigated the usage of simplified and convenient reduced models. The relationship between MATLAB and other simulation methods is explored in this paper. In addition, the uses of MATLAB-based program in material characterizations and image processing techniques are also discussed. More study on approaches, tools, and techniques will be required to improve the design and analysis of complicated processes for the manufacturing of real products.

Acknowledgements This work was supported by the Universiti Kuala Lumpur (grant number UniKL/CoRI/UER21014).

References

1. C. Trinh, D. Meimaroglou, S. Hoppe, Machine learning in chemical product engineering: the state of the art and a guide for newcomers. Processes **9**, 1456 (1–44)
2. E. Harley, G.R. Loftus, MATLAB and graphical user interfaces: Tools for experimental management. Behav. Res. Methods Ins. Comp. **32**, 290–296 (2000)
3. S.K. Dewangan, Importance & applications of digital image processing. Inter. J. Comp. Sci. Eng. Technol. **7**, 316–320 (2016)
4. H. Zhang, Z. Zhang, Z. Pei, Design and implementation of image processing system based on MATLAB, in *International Conference* (2015)
5. P.S. Duth, M.M. Deepa, Color detection in RGB-modeled Images using MATLAB. Int. J. Eng. Technol. **7**, 29–33 (2018)
6. V. Goel, S. Singhal, T. Jain, S. Kole, Specific color detection in images using RGB modelling in MATLAB. Int. J. Com. Appl. **161**, 38–42 (2017)
7. N. Sukamta, S.S. Sudarja, Characteristics of void fraction using image processing of two-phase flow of ali-pure water and glycerin (40–70%) on a transparent mini pipe with slope of 45oto the horizontal. J. Adv. Res. Exp. Fluid Mech. Heat Trans. **1**, 29–37 (2020)
8. E.A. Holm, R. Cohn, N. Gao et al., Overview: computer vision and machine learning for microstructural characterization and analysis. Metall. Mater. Trans. A **51**, 5985–5999 (2020)
9. J.D. Ortega, I.R. Vazquez, P. Vorobieff et al., A simple and fast matlab-based particle size distribution analysis tool. Int. J. Comp. Meth. Exp. Meas. **9**(4), 352–364 (2021)
10. P. Soille, *Morphological Image Analysis*, 2nd edn. (Springer, Berlin, 2004)
11. A. De Backer, C. Domain, C. Becquart, et al., A model of defect cluster creation in fragmented cascades in metals based on morphological analysis. J. Phys. Condens. Matter **30**(40), 405701(1–24) (2018)
12. C. Schneider, W. Rasband, K. Eliceiri, NIH Image to ImageJ: 25 years of image analysis. Nat. Methods **9**, 671–675 (2012)

Factors that Influence Sellers in Selection E-Marketplaces: A Systematic Literature Review

Eko Purwanto, Farahwahida Mohd, Zalizah Awang Long, and Singgih Purnomo

Abstract Electronic commerce is one of the implementations of information technology with the internet in the business sector. The rapid increase in e-commerce can significantly contribute to the economic sector. The e-marketplace is a new opportunity for online sellers to market and sell products without investing in a selling platform. This study aims to find literature on the factors influencing sellers in choosing e-marketplaces to sell their products. The guidelines used in this literature study are the selected reporting items for systematic review and meta-analysis guidelines, and the articles used are articles published in 2018–2022. There are 125 articles obtained from various databases, including IEEE Explore, Science Direct, and others. Validation and testing were conducted to obtain 36 articles that could be used as primary studies. As a result, ten factors influence sellers choosing an e-marketplace: platform, trust, service operations, marketing, and sales, quality of information, products, product reviews, perceived risk, ease of use, and payment channels.

Keywords Factor · Influence · Sellers · Selection · e-marketplace

E. Purwanto (✉) · F. Mohd · Z. A. Long
Universiti Kuala Lumpur Malaysian Institute Information Technology, Kuala Lumpur, Malaysia
e-mail: purwanto.eko@s.unikl.edu.my

F. Mohd
e-mail: farahwahidam@unikl.edu.my

Z. A. Long
e-mail: zalizah@unikl.edu.my

E. Purwanto
Faculty of Computer Science, Duta Bangsa University, Surakarta, Indonesia

S. Purnomo
Duta Bangsa University, Surakarta, Indonesia
e-mail: singgih_purnomo@udb.ac.id

© The Author(s), under exclusive license to Springer Nature Switzerland AG 2024 15
A. Ismail et al. (eds.), *Tech Horizons*,
SpringerBriefs in Applied Sciences and Technology,
https://doi.org/10.1007/978-3-031-63326-3_3

1 Introduction

Electronic commerce is one application of communication technology via the internet in the business sector. E-commerce is the activity of buying, selling, or trading goods through internet services [1]. According to Wood [2], it can positively impact the economy because of its flexibility and ability to create complete market access. Furthermore, [2] classifies the benefits of e-commerce as macroeconomic benefits and microeconomic benefits for individuals and groups. The development of e-commerce can make a significant contribution to the economic sector. One of the things that have caused the massive development of e-commerce is the increase in internet users due to the importance of the internet as the primary facilitator in e-commerce. The e-marketplace is a new opportunity for online sellers to market and sell products without the need to have their own sales platform or do not have to know legal aspects [3]. Some sellers offer many similar products on the e-marketplace, so sellers must be selective in choosing an e-marketplace as a place to market their products [4, 5]. Success in selecting e-marketplaces requires a key factor [6]. The e-marketplace selection model for online sellers can affect sales profits [7]. One of the functions of factor analysis is to form a model that helps sellers choose marketplaces to sell online. Selling online by utilizing the available marketplace platforms has several advantages, including market exposure, worldwide sales reach, and lower operating costs [8].

2 Methodology

Searching and selecting articles in systematic literature reviews must be carried out transparently. Meta-analyses are a guideline to increase transparency in systematic literature reviews [9].

This literature review is about what factors can influence sellers in choosing e-marketplaces. The stages in this study consist of the following steps: The foremost step is choosing an online database source as an iterative source in the research; forming keywords in the search process; defining general and specific criteria; pulling data; and reviewing and analyzing the results to answer study questions

2.1 Selecting Databases

The first step in this research is to select an online database source as a literature review. Online database sources are used to search for appropriate conferences, journals, and other publications. The following are selected sources in the systematic literature review in Google Scholar, IEEE Xplore Digital Library, Science Direct, and Springer Link.

2.2 Constructing Keywords

The keywords used in the search for articles are keyword combinations. The keyword combination in filtering data is to use Boolean operators such as OR and AND. Combinations will be used to lead to answers to research questions. The following varieties of keywords are used: (electronic marketplace, e-marketplace, or online marketplace) and (selection factors or influencing factors). It would help if one used a double apostrophe to avoid errors in the search because the search process is string data and consists of two words.

2.3 Inclusion and Exclusion Criteria

This research uses articles issued from 2018 to 2022 for inclusion considerations. The types of articles used are journal articles and conferences. However, in collecting information, consider other types of publications, such as articles on the website, as additional information or knowledge from sources that can be accounted for.

2.4 Data Extraction

The literature studied has been collected as many as 125 from all database sources. Then, from 125 articles, 68 articles were selected as prospect investigations. It is founded on the identification of the article and the concept. Then a review was carried out and produced 36 selected articles as material for this study. The process of selecting articles can be shown in Table 1.

Table 1 Number of studies in specified references

Data source	Stadies found	Candidate studies	Selected studies
Google Scholar	57	26	13
IEEE xplore digital library	23	17	9
Science direct	24	13	8
Springer link	21	12	6
Total	**125**	**68**	**36**

3   Result and Discussion

Many published papers related to e-commerce or e-marketplace were found. Some of these papers are from the Journal of Theoretical and Applied Electronic Commerce Research (#3), Electronic Markets (#2), Industrial Management and Data Systems (#2), International Conference on Management and Information Technology (#2), and others.

As in the collection of conference articles, it was found that the number of conference papers was 9 (25%), and journal articles were 27 (75%).

There are three major countries with the number of papers, namely Indonesia with 16 papers (44.44%), China with six papers (16.67%), and Australia with three papers (8.33%). Then the highest number of writers is in Indonesia, with 55 writers (52.38%), China, with 17 writers (16.19%); and Malaysia, with six writers (5.71%). In public, there are a total of 105 authors. In addition, some factors influence sellers in choosing e-marketplaces. Based on the literature study, ten factors influence sellers to choose e-marketplaces.

Established the literature review, it can be concluded that: (1) The platform, which consists of 15 articles (41.67%), is the most influential factor in choosing e-marketplaces. The e-marketplace platform is an important thing that sellers need to pay attention to in choosing an e-marketplace to sell their products [6, 10–23]. In selecting an e-marketplace, aside from considering the platform as the most influential factor, trust and service operations are also important factors for sellers in choosing an e-marketplace. (2) Trust comprises 13 articles (36.11%) in selecting e-marketplaces. Trust is also an important thing that must be considered because the transaction process in e-marketplaces is carried out virtually between sellers and buyers [6, 10–15, 19, 21, 24–27]. (3) Service operation consists of 10 articles (10.27%). In addition to platforms and trust, service operations are critical in selecting sellers' e-marketplaces. The transaction processes in e-marketplaces occur in real time, so e-marketplace platforms must provide good and quality operational services. (4) Marketing and sales have nine articles (25.00%). Choosing an e-marketplace to sell products, the seller also pays attention to marketing and sales factors because this factor is also the key to success in selling products through an e-marketplace [6, 10, 14, 15, 28–32]. (5) Information quality has five articles (13.89%). The selection of e-marketplaces needs to focus on the quality factor of the knowledge provided in the e-marketplace because good quality information for both customers and sellers will help in the transaction process through the e-marketplace [10, 30, 33–35]. (6) Product has four articles (11.11%). This factor is significant in choosing an e-marketplace because the seller always finds out what products are the most valuable sellers to sell on the e-marketplace [6, 10, 12, 21]. (7) Product reviews there are four articles (11.11%). In addition to the best-selling products on e-marketplaces, sellers always pay attention to product reviews when choosing an e-marketplace to sell their products [4, 34, 36, 37]. (8) Perceived risk is four articles (11.11%). Online buying and selling transactions involve sellers and buyers who do not meet face to face, so that they can increase the risk of loss due to transactions

Fig. 1 Proposed model

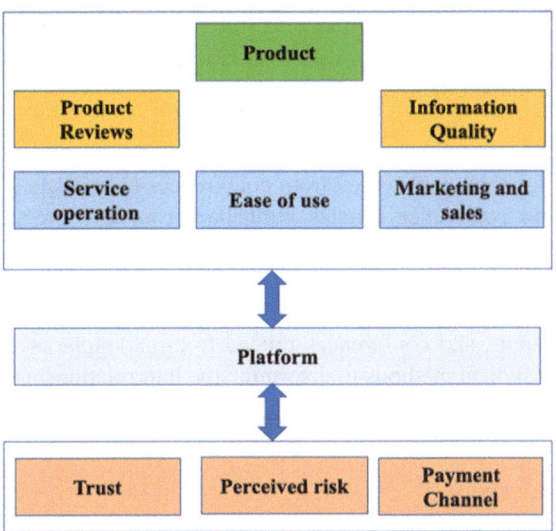

through third parties, namely e-commerce [38–41]. (9) Ease of use there are three articles (8.33%). Choosing an e-marketplace needs to ensure that the system in the e-marketplace is easy for sellers to use and understand so that it will increase seller satisfaction in using the e-marketplace [18, 33, 42]. (10) Payment channel has three articles (8.33%). The many choices of payment channels provided by the marketplace can reach more buyers and make it easier for sellers because they do not have to create multiple accounts to receive payments [6, 10, 14]. After reviewing the factors influencing e-marketplace choice from the seller's perspective, we can propose an e-marketplace selection model based on these factors, as displayed in Fig. 1.

4 Conclusion

Research on e-marketplaces is a multi-disciplinary understanding theme. There are several connected studies, such as management, commerce, and engineering. However, this research focuses on e-marketplaces, especially on the Scheme theme. This research found ten factors influencing sellers choosing e-marketplaces to sell their products. The ten factors are (1) platform; (2) trusts; (3) service operations; (4) marketing and sales; (5) information quality; (6) products; (7) product reviews; (8) perceived risk; (9) ease of use; and (10) payment channels. The platform is the most critical factor because the platform is an essential part of the e-marketplace.

There are two critical benefits to this research. The foremost benefit of this research is a hypothesis, which can be additionally used as a connection source in e-marketplaces. The second benefit is functional. This research can be utilized to determine factors influencing sellers in choosing e-marketplaces to market their products. The platform is the most critical factor for sellers in choosing an e-marketplace to market their products.

This analysis contributes to understanding the factors that impact sellers in choosing e-marketplaces. However, this research has restrictions, such as the number of databases accessed from conferences or journals. Then this research uses journal and conference articles published from 2018 to 2018. Then it is necessary to conduct trials using statistics to ensure that these factors influence sellers in choosing e-marketplaces to market their products.

For additional studies, it is essential to reproduce and extend the number of journal articles and conferences related to e-marketplaces. Then further research can use analytical methods to determine the interrelationships between factors.

References

1. E. Turban, J.K. Lee, D. King et al., *Electronic Commerce: A Managerial Perspective 2008*, 9th edn. (Springer, Berlin, 2018)
2. C.M. Wood, Marketing and e-commerce as tools of development in the Asia-Pacific region: a dual path. Int. Mark Rev. 301–320 (2004)
3. A. Kawa, LogForum. **15**, 521–529 (2019)
4. M. Arif, J.E. Suseno, R.R. Isnanto, Multi-criteria decision making with the VIKOR and SMARTER methods for optimal seller selection from several E-marketplaces. E3S Web Conf. **202**, 1–10 (2020). https://doi.org/10.1051/e3sconf/202020214002
5. L. Marlinda, Y. Durachman, E. Zuraidah, et al., Selection of E-marketplaces factors affecting Indonesian women's business loyalty using simple multi-attribute rating technique (SMART) method, in *2020 8th International Conference on Cyber IT Service Management (CITSM 2020)* (2020). https://doi.org/10.1109/CITSM50537.2020.9268823
6. P.Y. Meyliana, A.N. Hidayanto, H. Prabowo, The key success factors in e-marketplace implementation: a systematic literature review, in *Proceedings of 2018 International Conference on Information Management and Technology ICIMTech 2018*, pp. 443–448 (2018). https://doi.org/10.1109/ICIMTech.2018.8528189
7. T.M. Rofin, B. Mahanty, *Fulfillment mode selection for Indian online sellers under free and flat rate shipping policies* (Springer, US, 2021)
8. R. Rerung, *E-commerce* (Deepublish, Yogyakarta, 2018)
9. M.J. Page, D. Moher, Evaluations of the uptake and impact of the preferred reporting items for systematic reviews and meta-analyses (PRISMA) statement and extensions: a scoping review. Syst. Rev. **6**, 1–14 (2017). https://doi.org/10.1186/s13643-017-0663-8
10. J. Hatammimi, S.D. Purnama, Factors affecting prospective entrepreneurs to utilize e-marketplace: a study of business school students in Indonesia. J. Res. Bus. **11**, 1–11 (2022)
11. C.H. Park, Y.G. Kim, Identifying key factors affecting consumer purchase behavior in an online shopping context. Int. J. Retail Distrib. Manag. **31**, 16–29 (2003) https://doi.org/10.1108/09590550310457818
12. W.K. Chong, K.L. Man, M. Kim, The impact of e-marketing orientation on performance in Asian SMEs: a B2B perspective. Enterp. Inf. Syst. **12**, 4–18 (2018). https://doi.org/10.1080/17517575.2016.1177205
13. V. Kumar, O.G. Ayodeji, E-retail factors for customer activation and retention: an empirical study from Indian e-commerce customers. J. Retail. Consum. Serv. **59**, 102399 (2021). https://doi.org/10.1016/j.jretconser.2020.102399

14. V. Svatosova, The importance of online shopping behavior in the strategic management of e-commerce competitiveness. J. Compet. **12**, 143–160 (2020). https://doi.org/10.7441/joc.2020.04.09

15. L. Abdullah, R. Ramli, H.O. Bakodah, M. Othman, Developing a causal relationship among factors of e-commerce: a decision making approach. J. King Saud Univ. Comput. Inf. Sci. **32**, 1194–1201 (2020). https://doi.org/10.1016/j.jksuci.2019.01.002

16. A.A. Al-Tit, E-commerce drivers and barriers and their impact on e-customer loyalty in small and medium-sized enterprises (Smes). Bus Theory Pract. **21**, 146–157 (2020). https://doi.org/10.3846/btp.2020.11612

17. Y. Huang, G. Song, Q. Ye, Consumers' perceived trust evaluation of cross-border E-commerce platforms in the context of socialization. Proc. Comput. Sci. **199**, 548–555 (2021). https://doi.org/10.1016/j.procs.2022.01.067

18. H.N. Tran, M.D. Nguyen, Customer perception toward electronic commerce systems in Vietnam. Manag. Sci. Lett. **10**, 2861–2868 (2020). https://doi.org/10.5267/j.msl.2020.4.022

19. W. Thitimajshima, V. Esichaikul, D. Krairit, A framework to identify factors affecting the performance of third-party B2B e-marketplaces: a seller's perspective. Electron. Mark. **28**, 129–147 (2018). https://doi.org/10.1007/s12525-017-0256-3

20. H. Hallikainen, T. Laukkanen, National culture and consumer trust in e-commerce. Int. J. Inf. Manage. **38**, 97–106 (2018). https://doi.org/10.1016/j.ijinfomgt.2017.07.002

21. V. Babenko, Z. Kulczyk, I. Perevosova et al., Factors of the development of international e-commerce under the conditions of globalization. SHS Web Conf. **65**, 04016 (2019). https://doi.org/10.1051/shsconf/20196504016

22. J.H. Kim, Imperative challenge for luxury brands: generation Y consumers' perceptions of luxury fashion brands' e-commerce sites. Int. J. Retail Distrib. Manag. **47**, 220–244 (2019). https://doi.org/10.1108/IJRDM-06-2017-0128

23. M. Pan, R. Huang, M. Chi, S. Hu, The impact of platform flexibility and controls on platform attractiveness: an empirical study from the seller's perspective. Ind. Manag. Data Syst. **122**, 796–818 (2022). https://doi.org/10.1108/IMDS-08-2021-0528

24. L. Marlinda, S. Rusiyati, W.T. Adi et al., Selection of factors affecting women's loyalty in buying goods in indonesian e-marketplaces using the profile machine method. J. Theor. Appl. Inf. Technol. **97**, 2166–2178 (2018)

25. Y.Y. Chang, S.C. Lin, D.C. Yen, J.W. Hung, The trust model of enterprise purchasing for B2B e-marketplaces. Comput. Stand. Interfaces **70**, 103422 (2020). https://doi.org/10.1016/j.csi.2020.103422

26. L. Xiao, Analyzing consumer online group buying motivations: an interpretive structural modeling approach. Telemat. Inf. **35**, 629–642 (2018). https://doi.org/10.1016/j.tele.2018.01.010

27. M. Falahat, Y.Y. Lee, Y.C. Foo, C.E. Chia, A model for consumer trust in E-commerce. Asian Acad. Manag. J. **24**, 93–109 (2019). https://doi.org/10.21315/aamj2019.24.s2.7

28. A.M. Candraputri, A.C. Gunawan, F. Abdussalam et al., Analysis of factors that are affecting to customer loyalty in online to offline E-commerce case study at XYZ E-marketplace. Int. J. Adv. Sci. Technol. **29**, 5549–5567 (2020)

29. A.L. Harahap, S. Perdana, Analysis of employee performance appraisal using the BARS behavioral anchor rating scale method and management by objectives MBO at CV brilliant. IKRA-ITH Hum. J. Sos dan Hum. **5**, 18–26 (2021)

30. G.J.A. Santoso, T.A. Napitupulu, Factors affecting seller loyalty in business emarketplace: a case of Indonesia. J. Theor. Appl. Inf. Technol. **96**, 162–171 (2018)

31. S.M. Far, A model of the factors influencing consumer usage of B2C/C2C E-marketplaces Khalid ALAZAB page No: 2350. XII:2350–2363 (2020)

32. K. Kang, Strategic orientation, integrated marketing communication, and relational performance in E-commerce brands: evidence from Japanese consumers' perception. Bus. Commun. Res. Pract. **4**, 28–40 (2021). https://doi.org/10.22682/bcrp.2021.4.1.28

33. A.M. Sundjaja, D. Shukurnianto, A.P. Rulvi, R.H. Putra, Factors affecting buyer satisfaction of coffee beans at the online marketplace in Indonesia. Proc. 3rd Int. Conf. Inform. Multimedia Cyber Inf. Syst. ICIMCIS **2021**, 155–161 (2021). https://doi.org/10.1109/ICIMCIS53775.2021.9699202

34. L. Dennis, F. Ramdhana, T.C.E. Faustine, R.B. Hendijani, Influence of online reviews and ratings on the purchase intentions of gen Y consumers: the case of Tokopedia. Int. J. Manag. **11**, 26–40 (2020). https://doi.org/10.34218/IJM.11.6.2020.003

35. W. Wandoko, I.E. Panggati, The influence of digital influencer, e-WOM and information quality on customer repurchase intention toward online shop in e-Marketplace during pandemic COVID-19: the mediation effect of customer trust. J. Relatsh Mark. **21**, 148–167 (2022). https://doi.org/10.1080/15332667.2022.2035198

36. L.F.P.W. Fadhilah, L. Affifatusholihah, Factors affecting trust on purchase decisions through E-Marketplace. Int. J. Econ. Bus. Account. Res. **5**, 1120–1129 (2021)

37. E. Halim, V. Cornelya, H. Hartono, et al., Customer impulsive buying behaviors in Indonesia E-marketplace. Proc. 2022 Int. Conf. Inf. Manag. Technol. ICIMTech 2022, 533–538 (2022). https://doi.org/10.1109/ICIMTech55957.2022.9915065

38. S. Sfenrianto, T. Wijaya, G. Wang, Assessing the buyer trust and satisfaction factors in the E-marketplace. J. Theor. Appl. Electron. Commer. Res. **13**, 43–57 (2018). https://doi.org/10.4067/S0718-18762018000200105

39. D.A. Kindagen, H. Karamoy, R.T. Saerang, Perceived risk, trust, and purchase intention in online marketplace: perspective of consumers in Manado. Indonesia **9**, 715–725 (2021)

40. I Tzavlopoulos, K. Gotzamani, A. Andronikidis, C. Vassiliadis, Determining the impact of e-commerce quality on customers' perceived risk, satisfaction, value and loyalty. Int. J. Qual. Serv. Sci. **11**, 576–587 (2019). https://doi.org/10.1108/IJQSS-03-2019-0047

41. K. Wei, Y. Li, Y. Zha, J. Ma, Trust, risk and transaction intention in consumer-to-consumer e-marketplaces: an empirical comparison between buyers' and sellers' perspectives. Ind. Manag. Data Syst. **119**, 331–350 (2019). https://doi.org/10.1108/IMDS-10-2017-0489

42. M. Andarwati, P. Assih, F. Amrullah, et al., Success of small and medium enterprices (SMEs): actual technology use in e-Marketplace based on technology acceptance model (TAM) analysis, in *Proceedings of 2020 6th International Conference on Education and Technology, ICET 2020*, pp. 142–147 (2020). https://doi.org/10.1109/ICET51153.2020.9276594

Tax Data Analytics

Ahmad Faisal Hayek and Nora Azima Noordin

Abstract This study examines the application of data analytics in tax administration. The paper describes how data analytics methods such as predictive modeling, data mining, and machine learning have altered the way tax authorities operate by enhancing efficiency and accuracy while decreasing the amount of time and resources required for tax compliance. In addition, the article investigates how big data analytics has enabled tax authorities to scan massive volumes of data, including unstructured data in order to discover potential noncompliance and assess tax risks. The conclusion of the study is that the application of data analytics in tax administration has revolutionized tax administration by increasing compliance, decreasing expenses, and raising overall efficiency. It is anticipated that, as technology continues to improve, the application of data analytics in tax will continue to evolve, thereby boosting the efficiency of tax administration.

Keywords Tax analytics · Tax data analytics · Data analytics

1 Introduction

The benefits of data analytics extend beyond systems, departments, and organizational divisions to the entirety of an enterprise. Eventually, firms will be able to distribute important investment resources more quickly and confidently if they apply data and analytics to every element of their operations, from supply chains to tax systems. Today, not all firms have access to precise, exhaustive, and comprehensive tax data. To date, technological advances are bringing the accounting profession closer to being able to view all of their tax data. The big data revolution impacts

A. F. Hayek · N. A. Noordin (✉)
Faculty of Business, Higher Colleges of Technology, Sharjah Women's Campus, Sharjah, United Arab Emirates
e-mail: nazima@hct.ac.ae

A. F. Hayek
e-mail: ahayek@hct.ac.ae

© The Author(s), under exclusive license to Springer Nature Switzerland AG 2024   23
A. Ismail et al. (eds.), *Tech Horizons*,
SpringerBriefs in Applied Sciences and Technology,
https://doi.org/10.1007/978-3-031-63326-3_4

the accounting profession, particularly taxation. The greater collaboration between tax professionals and individuals with analytical backgrounds is one reason why the use of data analytics in tax has developed so significantly over the past two years. The successful implementation of data analytics in tax requires the merging of two worlds (tax and IT). The tax function is one of the most data-intensive roles within an organization, receiving trial balance data from finance systems as well as data from other transactional systems.

According to [12], tax functions spend more than 50% of their time getting tax data and less than 30% of their time on strategic tax analysis. Tax functions struggle to collect accurate and timely data, limiting their ability to contribute more strategically to enterprise-wide decisions. Being a downstream user of data, tax function is unable to tackle the problem of acquiring tax-ready information on its own, according to many tax executives.

1.1 Tax Data Analytics

The burgeoning field of data analytics in the taxation sector is known as tax analytics. It does this by applying the capabilities of emerging data analytics technologies to massive amounts of tax data. Some of these technologies include visualization, integration, and data fabric. The tax department's technical expertise is then put to use to provide significant insights and a deeper understanding, with the end goals of improving business performance and driving strategy through the adoption of more informed and timely decisions. The increased use of data analytics by tax authorities requires a mental shift on the part of businesses, specifically within their tax and finance departments, regarding how they collect, store, and evaluate data pertaining to taxes and finances. This shift is necessitated by the fact that tax authorities are increasingly using data analytics [3].

The following are five reasons why taxation activities may benefit from data analytics:

1. **Complete data visualization**

   With analytics software, one can quickly obtain data regarding each tax transaction and the overall performance.

2. **Real-time decisions**

   With analytics, there is no need to wait until the end of the month to generate reports and manually examine the data.

3. **Reply authoritatively to the tax authorities**

   In an increasingly digital tax environment, authorities expect one to be knowledgeable. If they want information, one will be able to respond confidently.

4. **Handle risk**

Data can center on any concerns with a particular customer, supplier, country, jurisdiction, or product or service.

5. **Always be ready for an audit**

Not only will a leading analytics platform help to remain compliant, but it will also store records of all transactions and validations that can be shared with tax authorities as needed.

1.2   Countries' with Tax Analytics Experience

According to [1] on the website of the Inter-American Center for Tax Administration (CIAT), the United States Internal Revenue Service (IRS) uses big data to combat tax fraud. One of the applied tactics, social media data mining, effectively recovers billions of dollars per year in missing taxes by demonstrating that people are living a more prosperous lifestyle than their tax records show. Big data solves this issue by using data classification and trail-based pattern recognition to distinguish between tax cheats and legitimate taxpayers. The Internal Revenue Service is investing in big data analytics and receiving a return on its investment. In a recent report, for instance, the IRS's Criminal Investigation Division stated that, despite significant workforce reductions, it had identified approximately 400% more tax fraud than the previous year and over 1000% more proceeds from other financial crimes than the previous year [4]. The objective of these tools is to improve case selection and coordination between IRS divisions. In the long run, data analytics and predictive policing will aid the IRS in identifying tax-reporting anomalies and tax evasion on a much larger scale.

In the UK, HM Revenue & Customs (HMRC) has developed the Connect system since 2017, a computerized data mining system of social network analysis software that cross-checks the tax records of companies and individuals with other databases to establish fraudulent activity. The software combines analytical tools and collects the information and implements predictive analysis similar to credit rating and has dynamic benchmarking. It seeks the correlation of income with lifestyle, comparing it with multivariate statistical models.

The Malaysian government has made attempts to utilize data analytics technologies and methodologies to enhance tax compliance and revenue collection. The establishment of the Malaysian Tax Assessment System (MTAS), an automated tax assessment system that leverages data analytics to identify tax risks and non-compliant taxpayers, is one of the important initiatives in this area. MTAS employs a combination of data mining, machine learning, and artificial intelligence to evaluate tax data and find trends and anomalies that may suggest potential tax evasion or noncompliance. This enables tax authorities to more precisely target their enforcement efforts and improve the accuracy of their tax assessments. According to the

Tax Division of the Ministry of Finance Malaysia, during the Malaysian government's 2017 budget announcement, the Minister of Finance established the Collection Intelligence Arrangement (CIA), which comprises the Companies Commis-
sion of Malaysia and two tax authorities, namely the Inland Revenue Board and
the Royal Malaysian Customs Department [9]. By facilitating data exchange and
matching between the CIA and international commitments with the Organization
for Economic Co-operation and Development (OECD) and Foreign Accounting Tax
Compliance Act (FATCA), big data analytics was applied. The use of data analytics
to support the Goods and Services Tax (GST) system in Malaysia is another such
example. In 2015, the Malaysian government established the Goods and Services
Tax (GST) system, which mandates businesses to collect and submit taxes on the
government's behalf. The Malaysian government has built a GST monitoring system
that uses data analytics to detect and prevent fraud and noncompliance in order to
facilitate compliance with the new tax regime. The application of data analytics in
tax administration has allowed the Malaysian government to increase tax compliance
and revenue collection while decreasing the burden on taxpayers. In order to ensure
that everyone pays their fair share of taxes, the government may detect tax risks
and non-compliant individuals more quickly and precisely by utilizing advanced
analytics tools and approaches.

The United Arab Emirates (UAE) has taken one of the most important steps in
this direction by establishing the Federal Tax Authority (FTA), which is in charge of
managing and collecting taxes inside the UAE. This is considered to be one of the
most important undertakings in this particular field. In order to support the FTA's
efforts in the administration of taxes, a number of data analytics technologies and
methods have been adopted. The FTA makes use of data analytics in order to keep
track of transactions and locate instances of possible tax evasion or non-compliance.
Additionally, the authority makes use of data analytics to monitor tax returns and
payments, recognize patterns, and forecast future revenue sources. This helps the
FTA to identify regions of possible tax risk and to target its enforcement activities
more effectively. The United Arab Emirates government has been able to improve
tax compliance and revenue collection while simultaneously lessening the burden on
tax payers as a result of the application of data analytics in tax administration.

2 Methodology

The approach of close reading has been embraced as a method of gaining insight from
a variety of previously published works as well as from a variety of other resources.
Journal articles, white papers, the content of websites, and blogs are considered in the
analysis. An analysis of the text has been carried out by investigating the intricacies
of the application of data analytics in the field of taxation. In order to do a more
in-depth text analysis, we first looked for patterns and categories of data analytics
that are related to taxation. After that, the patterns of data analytics were categorized
in relation to the efficacy with which they were being used in taxation.

3 Results and Discussion

This section discusses discoveries regarding the application of data analytics to various taxation tasks. It also discusses the implications of applying data analytics to various taxation challenges, such as tax administration.

3.1 Tax Administrations Using Big Data

Big data analytics is helping tax authorities detect non-compliance and fraud. Big data is huge, complicated data that is hard to process using typical methods. Tax administrations can now use big data to acquire insights into taxpayer behavior and patterns and find trends and anomalies that might assist them uncover non-compliance. Additionally, this learning is applied to computers that forecast likely feature events using historical data [6]. Tax administrations offer a way to identify abnormalities, which could help anticipate where they would recur. Online purchases, card transactions, billed offline purchases, and reactions to tax notices are all examples of data points in a taxpayer's tax journey that can be used for behavioral analytics, according to [2]. Collosa also said that when big data analytics and AI are coupled, authorities may significantly increase tax compliance. For example, past data can be used to predict how people will react when they get tax notices. Predictive analytics may also develop intricate risk profiles, study trends, spot potential audit problems, and flag higher-risk situations for additional investigation, thus blocking fraud avenues even before they begin. Additionally, these data may be submitted in different formats than how firms track and collect their own data.

According to [3], tax authorities are using real-time or near real-time data analytics engines to validate invoices and lag discrepancies, verify sales and purchase declarations, verify payroll and withholding declarations and compare data across jurisdictions and taxpayers. Based on these analyses, tax authorities make determinations, including tax and audit assessments. The most basic function of an intelligent data analysis platform should be to collect a large amount of tax data. After collecting relevant data, the platform should compare the data and set a standard value to predict tax crises. Data systems will issue stern reminders to people who exceed the standard value, allowing for the avoidance of some tax crisis issues. High standards exist for managers in tax management, where they must not only possess a solid professional background but also comprehend basic legal concepts related to taxation. Companies should pay more attention to professional abilities and cultivate and train more to increase the skills of employees since they will use numerous clever applications. In order to improve risk detection and optimize the performance of tax control tasks by focusing on the riskiest categories of taxpayers, one of the most significant ways to increase tax administration efficiency is based on strengthening the analytical function through the work of the Strategic Risk Department [8].

3.2   Data Mining Application in Taxation

Data mining is an application of data analytics in the world of taxation. Its objective is to examine tax data to identify trends and outliers in order to identify potential instances of tax evasion or noncompliance. Using statistical algorithms and machine learning techniques, data mining is the act of examining large volumes of data, particularly tax data, to uncover patterns and trends that may be suggestive of possible tax difficulties [7]. In the realm of taxation, data mining can be used to search for anomalies and outliers in tax data in order to identify probable cases of tax evasion or noncompliance. For example, a tax authority could use data mining to review tax data for taxpayers in a specific industry sector by analyzing the data provided by those taxpayers. Our research may find that a small fraction of taxpayers in this industry have declared unusually high or low amounts of revenue or expenditure. If they are discovered, their results could be unexpected. This information may be used by tax authorities to conduct additional investigations and identify potential instances of noncompliance with tax laws. In order to determine whether or not these taxpayers have engaged in fraudulent behavior, such as underreporting their income or misrepresenting their expenses, they may examine the financial records these taxpayers have maintained. Data mining can also be used to identify tax data patterns and trends that may be indicative of potential tax issues. This can be accomplished by evaluating the data in search of correlations between patterns and trends. For instance, a tax authority may use data mining to review tax data for taxpayers whose business structure has recently changed. This would be carried out for taxpaying individuals. This study may show that taxpayers who have recently reorganized their businesses are more likely to violate tax restrictions. This conclusion could be drawn based on the analysis's findings. The tax authority may use this information to build targeted outreach campaigns to educate these taxpayers about their tax duties and ensure their compliance with tax legislation. They could also use this information to detect future instances of noncompliance and more precisely target their enforcement efforts. If they had access to the aforementioned information, this would be doable.

3.3   Predictive Modeling Application in Taxation

Taxation uses predictive modeling, a sort of data analytics, to anticipate future income streams based on historical tax data. Using statistical algorithms and machine learning approaches, predictive modeling analyzes vast quantities of historical tax data and identifies patterns and trends. Predictive modeling can be used in taxation to determine which taxpayers are more likely to violate tax regulations. By evaluating past tax data, prediction algorithms can uncover anomalies and patterns suggestive of probable tax evasion or noncompliance. For instance, a predictive algorithm could determine that taxpayers who have recently altered their business structure or made substantial purchases are more likely to violate tax regulations. In addition to predicting future tax income, predictive modeling can also be used to estimate future tax expenditures. By evaluating historical tax data and economic variables, predictive algorithms can accurately project future tax collections. Tax authorities can use this information to plan their revenue collection activities and guarantee they have adequate resources to meet their budgetary needs. The Audit Division's Advanced Database System (ADS), a data warehouse created to support tax compliance applications like the Audit Select system, which used predictive models to identify candidates for sales tax audits, was one of the first to use predictive modeling in tax administration [10].

3.4 Descriptive Analytics Application in Taxation

Descriptive analytics can be used to develop models of questionable behavior or transactions. Multilevel association rules, multidimensional association rules, and quantitative association rules could be types of association rule analysis for the descriptive tasks [13]. Algorithms that create association rules describe potentially fraudulent scenarios. Cluster analysis aggregates data into subsets with linked patterns, or high-quality clusters with high intra-class similarity and low inter-class similarity [5]. In taxation, descriptive analytics is a type of data analytics used to identify behavioral tendencies among taxpayers. Descriptive analytics is the process of analyzing historical tax data to comprehend what has transpired in the past and to uncover patterns and trends that might inform future actions. In taxation, descriptive analytics can be used to uncover trends in taxpayer behavior by studying tax returns over time. Using descriptive analytics, a tax authority may study tax returns for a given industry sector over a five-year period. Descriptive analytics can be used to uncover trends in taxpayer behavior by studying tax returns for specific groups of taxpayers. A tax authority could, for instance, review the tax returns of high-income taxpayers over a five-year period using descriptive analytics. This investigation may reveal a large increase in the number of high-income taxpayers claiming charitable deductions. The tax authority might use this information to create targeted outreach campaigns to educate high-income people on the rules and regulations governing charitable donations. This information may also be used to discover potential tax law infractions, such as taxpayers claiming charitable deductions for gifts to bogus organizations.

3.5 Artificial Intelligence Application in Taxation

Using artificial intelligence (AI), global tax authorities are automating tax assessment and compliance processes. AI uses machine learning algorithms and natural language processing to examine huge amounts of data and make predictions or determinations based on that data. AI is a crucial instrument that may boost productivity and save expenses. It has also been shown to be an excellent technique for decreasing the risk of human error [11]. By evaluating tax returns and finding any flaws or omissions, AI can be utilized to automate the tax assessment process. An AI system may, for instance, scan a tax return and detect any items that appear to fall outside the normal range of values for the taxpayer's industry or income category. Also, the algorithm may identify items that contradict the taxpayer's previous tax returns. By automating the tax assessment process, AI can aid tax authorities in processing tax returns more efficiently, hence reducing the need for human review and involvement. This can reduce the administrative burden on tax authorities and free up resources to address more complex problems. AI can also be used to automate the tax compliance process as a result of analyzing taxpayer data and identifying probable instances of noncompliance. For instance, an AI system may review a taxpayer's financial data and flag any transactions that contradict the taxpayer's declared income.

4 Conclusion

In conclusion, the application of various types of data analytics in tax has altered the way tax authorities operate, boosting efficiency and accuracy while decreasing the amount of time and resources needed for tax compliance. The application of data analytics technologies such as predictive modeling, data mining, and machine learning has enabled tax authorities to recognize patterns and trends in taxpayer behavior, detect tax fraud and evasion, and assure tax law compliance. Moreover, the use of big data analytics can assist tax authorities in processing enormous quantities of data from diverse sources, such as social media, in order to detect potential fraudulent activities. This strategy can also assist tax authorities in identifying previously unknown fraud schemes, thereby improving their capacity to detect and prevent tax fraud. Overall, the application of data analytics in tax has transformed the operation of tax authorities by enhancing compliance, decreasing expenses, and enhancing overall efficiency. With the continuing development of technology, it is anticipated that the application of data analytics in tax will continue to expand and change, thereby boosting the efficiency of tax administration.

References

1. A. Collosa, Use of Big data in Tax Administrations. Inter-American Center of Tax Administrations. https://www.ciat.org/use-of-big-data-in-tax-administrations/ (2021a). Accessed 10 March 2023
2. A. Collosa, Big data in Tax Administrations. https://kluwertaxblog.com/2021/07/16/big-data-in-tax-administrations/ (2021b). Accessed 10 March 2023
3. EY Global, Ernst & Young Global Ltd.: How data analytics is transforming tax administration. https://www.ey.com/en_si/tax/how-data-analytics-is-transforming-tax-administration (2020)
4. J.B. Freeman, The IRS and Big data: the future of fighting tax fraud. Freeman Law. https://freemanlaw.com/the-irs-and-big-data-the-future-of-fighting-tax-fraud/ (2019)

5. R. Gupta, N. Singh, Prevention and detection of financial statement fraud—an implementation of data mining framework. Int. J. Adv. Comput. Sci. Appl. **50**(8), 7–14 (2012). https://doi.org/10.5120/7789-0889

6. A.Y. Hayek, Data science and external audit, in *Sustainable Development through Data Analytics and Innovation. Progress in IS*, edited by J. Marx Gómez, L.O. Yesufu, pp. 45–62 (Springer, Cham, 2022)

7. A.F. Hayek, N.A. Noordin, K. Hussainey, Machine learning and external auditor perception: an analysis for UAE external auditors using technology acceptance model. J. Account. Manag. Inf. Syst. **21**(4), 475–500 (2022). https://doi.org/10.24818/jamis.2022.04001

8. L. Lin, Application of big data model in financial taxation management. Sci. Program. 1–10 (2021). https://doi.org/10.1155/2021/7001456

9. B. Mahfuzah, Big data in taxation: towards better compliance & collection, tax division ministry of finance Malaysia, in *13th International Tax Administration Conference*, University of New South Wales, 5–6 April 2018

10. D. Micci-Barreca, S. Ramachandran, Improving tax administration with data mining. http://spsslietuva.com/media/collateral/modeling/tax.pdf (2004). Accessed 15 March 2023

11. N.A. Noordin, K. Hussainey, A.F. Hayek, The use of artificial intelligence and audit quality: an analysis from the perspectives of external auditors in the UAE. J. Risk Financ. Manag. **15**(8), 339 (2022). https://doi.org/10.3390/jrfm15080339

12. Tax Function of the Future series—Unlocking the power of data and analytics—Tax—PwC UK blogs. https://pwc.blogs.com/tax/2015/11/tax-function-of-the-future-series-unlocking-the-power-of-data-and-analytics-.html (2015). Accessed 15 March 2023

13. V. Thillainayagam, Data mining techniques and applications—a review. I-manager's J. Softw. Eng. **6**(3), 44–48 (2012). https://doi.org/10.26634/jse.6.3.1791

Digital Competence Assessment for Royal Malaysian Air Force Aircraft Maintenance Technicians

T. Nanthakumaran Thulasy, Istas Fahrurrazi Nusyirwan, Noorlizawati Abd Rahim, Astuty Amrin, Puteri N. E. Nohuddin, Zuraini Zainol, Nora Azima, and Lawal Yesufu

Abstract Aircraft mechanics need digital competencies for maintaining sophisticated, interactive, and collaborative aircraft engines and systems. The RMAF fundamental training does not assess the digital skills of technicians. Therefore, this study devised survey assessment components based on IR4.0 that incorporate digital skills and competencies for aviation maintenance tasks. This study employed seven criteria to evaluate the skills and competencies of aircraft maintenance technicians: problem

T. Nanthakumaran Thulasy (✉) · I. F. Nusyirwan · N. A. Rahim · A. Amrin
Razak Faculty of Technology and Informatics, Universiti Teknologi Malaysia, 54100 Kuala
Lumpur, Malaysia
e-mail: tnanthakumaran@graduate.utm.my

I. F. Nusyirwan
e-mail: istaz@utm.my

N. A. Rahim
e-mail: noorlizawati@utm.my

A. Amrin
e-mail: astuty@utm.my

P. N. E. Nohuddin · N. Azima
Higher Colleges of Technology, Sharjah, United Arab Emirates
e-mail: pnohuddin@hct.ac.ae; puteri.ivi@ukm.edu.my

N. Azima
e-mail: nazima@hct.ac.ae

P. N. E. Nohuddin
Institute of IR4.0, Universiti Kebangsaan Malaysia, 43000 Bangi, Malaysia

Z. Zainol
Faculty of Defence Science and Technology, Universiti Pertahanan Nasional Malaysia, Kuala
Lumpur, Malaysia
e-mail: zuraini@upnm.edu.my

L. Yesufu
Hult International Business School, Dubai Internet City, United Arab Emirates
e-mail: lawal.yesufu@faculty.hult.edu

© The Author(s), under exclusive license to Springer Nature Switzerland AG 2024   33
A. Ismail et al. (eds.), *Tech Horizons*,
SpringerBriefs in Applied Sciences and Technology,
https://doi.org/10.1007/978-3-031-63326-3_5

solving, communication, active learning, technical knowledge, analytical and critical thinking, technological skills and experience, and lessons learned. Three expert review panels on aircraft maintenance and IR4.0 readiness validated the developed survey. The study found that ITAS technicians had the lowest mean evaluation score, 15.441 and the highest standard deviation, 8.630, indicating a wide range of variance compared to other selected squadrons. The F-statistic and p-value of the ANOVA analysis indicate < 0.001 which significantly shows that assessment scores differed across groups. Thus, in line with IR 4.0 requirements, enhancement trainings are desirable to upskill the digital skills of technicians.

Keywords Aircraft maintenance · Assessment · Industrial revolution 4.0 · Digital competencies

1 Introduction

The Royal Malaysian Air Force (RMAF) has established its genuine capacity, credibility, and force as a national air force in terms of the efficacy of its air operations actions. Each year, the RMAF's engineering branch verifies that all aircraft maintenance technicians are adequately trained and competent to perform their duties. RMAF strictly provides excellent aircraft maintenance so as to maintain an aircraft airworthy and dependable. Aircraft mechanics are in charge of the upkeep and repair of airplanes and helicopters. Their competence encompasses aircraft maintenance, including engine systems, armament systems, flight controls, and electrical systems, as well as avionics systems [1]. The Institute Technology of Aerospace (ITAS) is an air force college-based training institute responsible for administering fundamental of aircraft technical and maintenance training programs.

As new aircraft or weapon systems are introduced, aircraft technicians must continuously upgrade their knowledge and skills to keep up with the latest technological advances [2]. Furthermore, the aerospace industry is characterized by rapid technological development and immense variety. Every 10–15 years, aircrafts are upgraded with enhanced technology and flying capabilities that can be tailored to the requirements of a particular organization, despite the fact that most aircrafts have a lifespan of 40 years [3]. Therefore, it is crucial for aviation industry professionals such as aircraft engineers and technicians to keep abreast of new technologies [4].

As a consequence of the digital transformation, technicians will routinely process large amounts of information and data. Eventually, artificial intelligence (AI) will enable humans and machines to collaborate effectively. Technicians will rely on intelligent devices and tablets as the primary means of communication and control for a variety of diagnostic equipment. New technicians must participate in process planning and optimization efforts [5].

Numerous occupations, such as those requiring expertise in diagnostics, machine problems, resource management (including human management and goal formulation), and so on, will become obsolete as automation and artificial intelligence become increasingly prevalent [6]. Due to the increasing diversity and complexity of digital skills, businesses need systematic methods for assessing current competencies and planning for the future.

Consequently, it is essential to evaluate the competence of aircraft maintenance specialists. In addition, the results of the skill gap analysis could assist RMAF training and education facilities in determining which IR 4.0 and aviation 4.0-related programs and courses to offer technicians. Therefore, modern aviation maintenance necessitates that aircraft maintenance personnel receive training and retraining in order to integrate new aircraft technologies and systems, particularly those of IR4.0 and aviation 4.0.

This study aims to develop an assessment instrument and ascertain the level of IR4.0-based skills possessed by aircraft maintenance technicians in the RMAF. IR4.0 readiness is dependent on the digital skills of each technician in performing aircraft maintenance tasks, such as aircraft maintenance information system, sensor troubleshooting duties, and avionics systems in preventive maintenance [7]. In addition, this research collects data on the digital skill profiles of technicians for determining their level of competency. The anticipated outcome at this stage is a validated set of survey questionnaires that will be used to assess the level of IR 4.0-based skills.

The paper is structured as follows: Sect. 2 elaborates the methodology of the instrument development and validation. Section 3 presents the result and discussion regarding the skills and competencies of aircraft maintenance technicians. Lastly, Sect. 4 is a conclusion of the paper.

2 Methodology

Research questionnaire is used to pursue dependable data acquisition. The data can be analyzed using descriptive approaches [8] and organizing sampling at a specific level based on parameters from earlier phases, which enhances the accuracy of the conclusions [9].

In this study, seven dimensions, namely problem solving, communication, active learning, technical knowledge, analytical and critical thinking, technological skills and experience, and lesson learned, were identified through a systematic literature review (SLR) in an earlier stage of this research for evaluating the skills and competence of aircraft maintenance technicians based on IR 4.0 environment [2, 7]. These dimensions are regarded as providing valuable scopes and boundaries for devising methods, instruments, or procedures for evaluating the acquisition, transmission, and development of an individual's technical skills and competency. Specifically, questionnaires are designed to assess digital competency relevant to these dimension.

Stage 1: Questionnaires Development

The first stage involves the questionnaires development. The scope of questionnaire comprises of the seven dimensions enumerated in Table 1. Each dimensions is comprised of four items (questions) that assess the digital competency of aircraft maintenance technicians, such as "I can APPLY problem resolutions using the latest digital tools that may assist me in problem solving" and "I can UNDERSTAND and communicate with other technicians when discussing aircraft rectification using digital test equipment."

Stage 2: Instrument Validation

Then, stage 2 is the process of instrument validation. Expert reviewer panels evaluated the questionnaires, gave feedbacks and queries on the items and examined the survey for common errors, such as ambiguous questions and linguistic issues. Authors of [10] suggested that evaluating the content validity is to ensure the novelty of instruments. Content validity is the process of ensuring that a new survey instrument includes all required elements and excludes unwanted items from the concept domain. The evaluative approach to content validity [11] begins with a literature review and evaluation by an expert panel. For the validation of questionnaires, researchers and experts discuss the content validity. Table 2 lists the three committees of experts that evaluated the instrument in this research.

The following steps in the validation process entails analyzing the survey based on the marks given by the panelists on the validation forms. The Context Validity Index (CVI) is a crucial instrument for determining whether or not a questionnaire has been adequately validated. Figure 1 depicts the validation outcomes for Item-CVIs of 0.78 or greater, as well as Scale-level CVI Universal Agreement (S-CVI/UA) and S-CVI/Average of 0.8 and 0.9 or greater.

Step 3: Validation Using Principal Component Analysis

Principal Component Analysis (PCA) is typically used by researchers to identify this questionnaire's primary components [12]. Factor loading denotes that objects must

Table 1 Questionnaires multidimensional construct and items

Assessment multidimensional	Items
Problem solving measures an aircraft maintenance technician's competence to maintenance and resolve digital system issues	4
Communication measures an aircraft maintenance technician's ability to interact with coworkers and share ideas using communication digital tools	4
Active learning measures an aircraft maintenance technician's ability to explore current technology aircraft maintenance fix	4
Technical knowledge is a measurement of an aircraft maintenance technician's ability to work with aircraft systems	4
Analytical and critical thinking Analytical and critical thinking measures an aircraft maintenance technician's capacity to simplify challenging discussions or produce new ideas	4
Technological skills is a measurement of an aircraft maintenance technician's ability to complete a task utilizing specialized knowledge	4
Experiences and lessons learned measures an aircraft maintenance technician's proficiency and ability to seek for more resources through knowledge bases	4

Table 2 Expert review selection profiles

Expert panels	Working place	Background
Professor	University Kuala Lumpur—(Malaysian Institute of Aviation Technology)	License aircraft engineer, aircraft maintenance engineering
Senior lecturer	Institute of IR 4.0, The National University of Malaysia	IR 4.0 readiness digitalization
Senior engineer	Central aerospace engineering services establishment, RMAF	Mechanical engineering—aeronautic

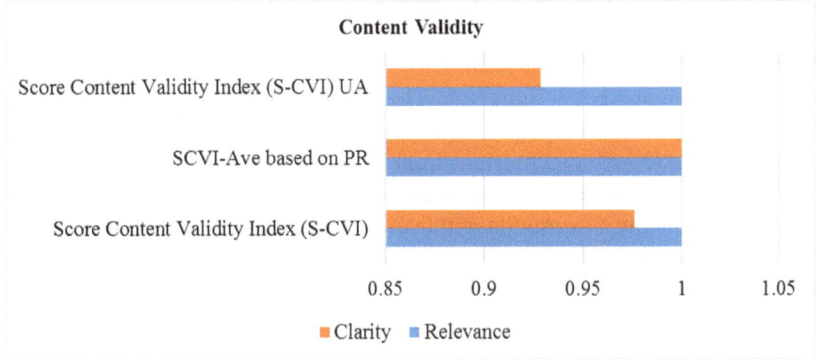

Fig. 1 Score context validity index of survey items

share the same factor, and an acceptable factor loading is typically greater than 0.6. In this study, PCA is also used to determine the inter-correlation between matrix variables. Variable intercorrelations are used to generate the main components in the second stage, which employs the correlation matrix. The results of PCA for all survey constructs created are summarized in Table 3, all factor loading for each component (RC1) exceeds 0.6.

Table 3 PCA and uniqueness values

Dimension	RC1	Uniqueness
PS	0.86	0.26
C	0.87	0.24
AL	0.82	0.32
TK	0.86	0.26
ACT	0.83	0.32
TS	0.87	0.24
ELL	0.91	0.16

3 Results and Discussion

The developed assessment instrument has been distributed to randomly selected aircraft maintenance technicians. This is aimed to discover the gaps in digital skills and competency among RMAF current aircraft maintenance technicians between squadrons. The sample size (N) for ITAS is 119 technicians, 39 technicians from No. 10 Squadron, 127 technicians from No. 12 Squadron, and 165 technicians from No. 8 Squadron. No. 8 Squadron has the largest sample size ($N = 165$), which suggests that the estimates of the mean and standard deviation for this group are likely to be more precise than for groups with smaller sample sizes; however, it is unavoidable due to availability of technicians at the bases.

Figure 2 and Table 4 show the mean, standard deviation, and sample size (N) for the variable of interest for each of the four groups: ITAS, No. 10 Squadron, No. 12 Squadron, and No. 8 Squadron. The mean score for technicians from ITAS center is 15.441, which means that, on average, is the lowest mean assessment score. Meanwhile mean score for technicians from No. 10 Squadron is 28.532, from No. 12 Squadron is 27.492, and from No. 8 Squadron is 26.805. The standard deviation (SD) quantifies the quantity of variation within each group. A smaller standard deviation indicates that the group's values are more securely concentrated around the mean, whereas a larger standard deviation indicates that the values are more dispersed. The highest standard deviation of 8.630, indicating that assessment score for digital skills has a wide range of values among ITAS technicians.

Fig. 2 Descriptive graph

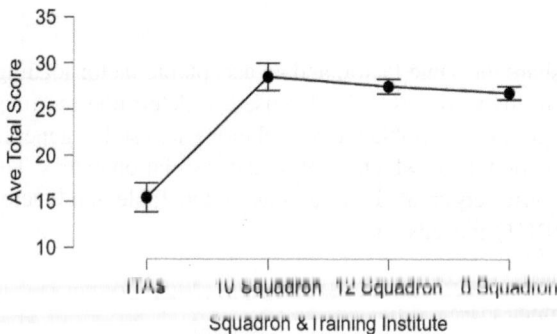

Table 4 Descriptive analysis

Center and squadrons	Mean	SD	N
ITAS	15.441	8.630	119
No. 10 squadron	28.532	4.574	39
No. 12 squadron	27.492	4.561	127
No. 8 squadron	26.805	4.650	165

Table 5 ANOVA variance analysis

ANOVA—Ave total score

Homogeneity correction	Cases	Sum of squares	df	Mean square	F	p
Welch	Squadrons and ITAS	12,355.52	3	4118.51	68.745	< 0.001
	Residuals	15,749.99	151.12	104.21		

Table 5 presents the ANOVA analysis which is used to calculate the homogeneity correction cases (sum of squares) and one for the within-group variability (the residuals). The sum of squares (SS) is 12,355.52 of variation in the data that can be attributed to the differences between the groups being compared and the residual value is 15,749.99. The higher SS value indicates more variation between technicians' assessment scores from the center and squadrons. The degrees of freedom (df) is 3, which indicates that there were 4 groups being compared and the df for the within-group SS is 151.

The mean square (MS) for each source of variability is obtained by dividing the SS by the df. In this case, the mean square for the between-group variability is 4118.506 (12,355.52 divided by 3) and the mean square for the within-group variability is 104.211 (15,749.99 divided by 151). Finally, the F-statistic is 68.745 (4118.506 divided by 104.211). The p-value associated with this F-statistic is less than 0.001, which indicates that the differences between the groups are statistically significant.

In summary, using an assessment instrument, the analysis reveals that No. 10 Squadron technicians are anticipated to be more experienced than those of other squadrons because they operate more advanced aircrafts. ITAS center technicians have the lowest mean assessment score and the highest standard deviation, indicating a wide range of values. The F-statistic and p-value from the ANOVA analysis indicate that there is a significant difference in assessment scores between ITAS and squadrons. The study suggests additional training and development programs to help technicians upskill the digital skills and competency gap.

4 Conclusion

The most recent paradigm for identifying technology related skills is digital competence. Thus, ICT skills, technology skills, information technology skills, twenty-first century skills, information literacy, digital literacy, and digital skills have been used to designate the skills and competencies necessary for using digital technologies [13]. Therefore, it is essential in the IR 4.0 era to evaluate the skills and competencies of the personnel in order to operate and administer ICT-related systems and devices.

Therefore, the purpose of this study was to construct and adapt assessment components that incorporate digital skills and competencies for performing aircraft maintenance duties based on IR 4.0. RMAF aircraft maintenance technicians' digital skills and competencies are assessed across squadrons. From the study, ITAS technicians had the lowest mean evaluation score and the highest standard deviation, suggesting a broad range of values. The F-statistic and p-value of the ANOVA analysis suggest that assessment scores varied between groups. To improve technicians' digital abilities, the finding suggests further digital competency training.

References

1. T.N. Thulasy, P.N.Nohuddin, I.F. Nusyirwan, N.A. Rahim, A. Amrin, Skill set issues in aircraft maintenance from industrial revolution 4.0 context: a document analytics survey. Hum. Syst. Manag. 1–11 (2021). https://doi.org/10.3233/HSM-210013
2. T. Güneş, U. Turhan, B. Açıkel, An assessment of aircraft maintenance technician competency. Int. J. Avi. Sci. Tech. 22–29 (2020)
3. I.A. IATA: *Aircraft Technology Roadmap to 2050*. International Air Transport Association, Geneva (2020)
4. Skills Future.sg: *Skills Framework for Aerospace: A Guide to Occupations and Skills*. https://www.nlb.gov.sg/biblio/203165741 (2017). Accessed 22 Feb 2023
5. M. Lee, J.J. Yun, A. Pyka, D. Won, F. Kodama, G. Schiuma, H. Park, J. Jeon, K. Park, K. Jung, M.-R. Yan, S. Lee, X. Zhao, How to respond to the fourth industrial revolution, or the second information technology revolution? Dynamic new combinations between technology, market, and society through open innovation. J. Open Inn. Tech. M. Complex. **4**(3), 21 (2018)
6. World Economic Forum: Education and Skills 2.0: New targets and innovative approaches. http://www3.weforum.org/docs/GAC/2014/WEF_GAC_EducationSkills_TargetsInnovativeApproaches_Book_2014.pdf (2014). Accessed 22 Feb 2023
7. T.N. Thulasy, P.N. Nohuddin, I.F. Nusyirwan, N.A. Rahim, A. Amrin, S. Chua, Skills assessment criteria for aircraft maintenance technician in the context of industrial revolution 4.0. J. Aerosp. Technol. Manag. 1–14 (2022). https://doi.org/10.1590/jatm.v14.1286
8. N. Nimvyap, L.F. Miba'am Walwai Benjamin, The prospects and challenges of entrepreneurship and venture creation among graduates of tertiary institutions in plateau state Nigeria. Int. J. Manag. Stdies. Bus Entrep. Res. **5**(2), 1–14 (2020)
9. A. Mallik, M. Banerjee, G. Michailidis, M-estimation in multistage sampling procedures. Sankhya A **82**(2), 261–309 (2020). https://doi.org/10.1007/s13171-019-00194-2
10. H. Taherdoost, Sampling methods in research methodology; how to choose a sampling technique for research; how to choose a sampling technique for research. Int. J. Acad. Res. Manag. **5**(2), 18–27 (2016). ISSN: 2296-1747
11. V. Quiroz, D. Reinero, P. Hernández, J. Contreras, R. Vernal, P. Carvajal, Development of a self-report questionnaire designed for population-based surveillance of gingivitis in adolescents: assessment of content validity and reliability. J. Appl. Oral Sci. **25**(4), 404–411 (2017). https://doi.org/10.1590/1678-7757-2016-0511
12. K.L. Sainani, Introduction to principal components analysis. PM&R **6**(3), 275–278 (2014). https://doi.org/10.1016/j.pmrj.2014.02.001
13. L. Ilomäki, A. Kantosalo, M. Lakkala, What is digital competence? in *Linked Portal*. European Schoolnet, Brussels. http://linked.eun.org/web/guest/in-depth3 (2011). Accessed 13 Mar 2023

IoT-Based Wearable Device for Position Tracking and Visualization

Syadrizzad Syarifuddin Yusoff, Noor Hidayah Mohd Yunus, Jahariah Sampe, and Hanani Nadzirin

Abstract The Internet of Things (IoT) technology has received a lot of attention in business, smart cities, smart grids, autonomous vehicles, industrial internet and academic worlds in recent years. This gives the idea to make a tracking system with IoT applications that consist of real-time visualization where it can monitor and track a person using the device. Nowadays, every parent allows and trains their children to explore everywhere, for example, playing with friends in the playground, going to school and even to the nearby shop to buy some necessities. However, this situation often makes parents worry about their children's safety when they are outside the residential area. As reported by Bukit Aman's criminal investigation department (CID), an average of 78 cases of abduction and disappearance of children under the age of 12 were reported every month in 2022. Therefore, this project aims to assist parents in tracking the location of their children's position also used to be a smart tracker for visualization detection and could track the location of vehicles. The main components used are the ESP32 microcontroller as the integrated wireless communication and the Global Positioning System (GPS) as the geolocation and time provider. This proposed project can help parents monitor the whereabouts of their children through the internet and keep track of the children in real time.

S. S. Yusoff · N. H. M. Yunus (✉)
Advanced Telecommunication Technology, Communication Technology, Section, Universiti Kuala Lumpur British Malaysian Institute, Batu 8, Jalan Sungai Pusu, 53100 Gombak, Selangor, Malaysia
e-mail: noorhidayahm@unikl.edu.my

S. S. Yusoff
e-mail: syadrizzad.yusoff@s.unikl.edu.my

J. Sampe
Institute of Microengineering and Nanoelectronics, Universiti Kebangsaan Malaysia, 43600 Bangi, Selangor, Malaysia
e-mail: jahariah@ukm.edu.my

H. Nadzirin
Water and Energy Section, Universiti Kuala Lumpur Malaysian France Institute, Bandar Baru Bangi, Selangor, Malaysia
e-mail: hanani@unikl.edu.my

Keywords Wearable device · GPS · Location tracker · ESP32 Cam · Arduino IDE

1 Introduction

Nowadays, most parents are too busy with their careers and consume too little time to spend with their children and have to leave the children at daycare when they go to work. However, the safety and whereabouts of children must be supervised and taken seriously. Children are the most important members of the family to be given attention to and would be the responsibility of parents or guardians to look after and monitor their whereabouts.

The development of data technology promotes innovation and enhancement through the development of technology experiments for the care of children and family members. In this development, information technology using IoT is implemented in applications such as process monitoring, place tracking, and assistance for security purposes. In the technological era of industrial revolution 4.0 (IR 4.0), IoT application aims to assist and ease all aspects of human life [1–3].

A wearable monitoring device with an integrated global positioning system (GPS) and wireless communication is one of the easiest ways to track the user's location which is attached to the tracking device [4, 5]. GPS is a satellite-based navigation system that transmits data from Earth-orbiting satellites to GPS receivers on the ground [6, 7]. Data from various satellites are received by GPS receiver in the National Marine Electronics Association (NMEA) protocol form [8]. The data received in NMEA code contains a combination of information such as the user's current position and time on Earth.

In this paper, the proposed position tracking system uses IoT technology and adopts the ESP32 microcontroller integrated camera through a geolocation technique to inform the parents or guardians about the location of the family member or children and get a person's visualization easily. The proposed system is further explained in the following section.

2 Methodology

Figure 1 shows the block diagram of the tracking system. On the front end of the wearable device side, the ESP32 microcontroller is used for the operation system and the ESP32 camera is connected to record the visualization of the device. On the back end of the monitoring system side, the smartphone is used to view the user's recorded position that is attached to the tracking device.

The ESP32 microcontroller is linked to the GPS and the antenna module in the serial connection. The GPS antenna of the device is connected to a GPS satellite. Then, the receiver antenna of the GPS sends data in terms of coordinates and period to the microcontroller. Then, the ESP32 microcontroller instructs the GPS module

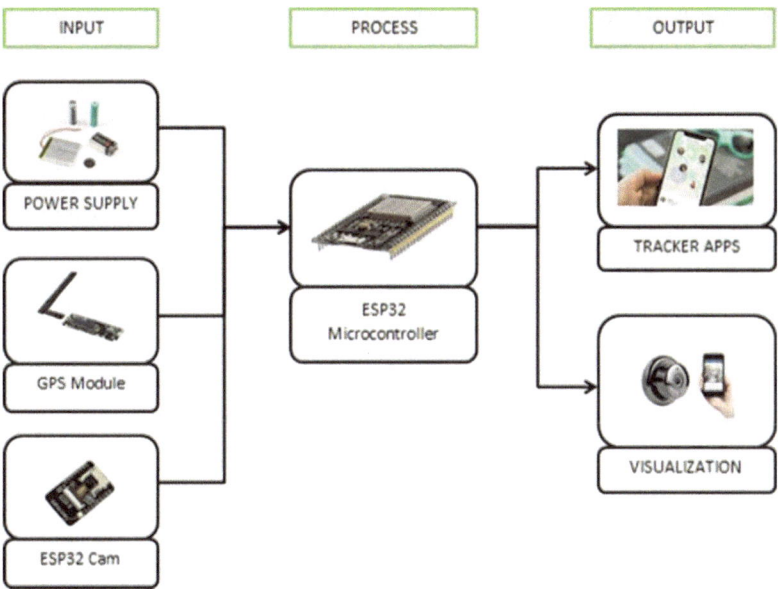

Fig. 1 Block diagram of the system

to send the data wirelessly to the enabled smartphone device. When the data from the GPS is found, concurrently the ESP32 camera also transmits the visualization recording from the front-end device to the tracker apps that use Blynk. The information data from the ESP32 and GPS antenna are displayed on the smartphone through the Blink app as well as on the serial monitor.

Figure 2 shows the operational flow of the system. The process starts with setting the radius limit up to 500 m and initializing the GPS to detect the device's location. When the GPS is found, the system is ready to operate and send to the ESP32 microcontroller to process the data. But at the same time, the camera is initialized for visualization records to the ESP32 microcontroller. When the user of the front-end device starts to make a move, the system can read the position and display the view of the location. The notification and the information data will be shown in tracker apps which display the GPS coordinate of the device and the visualization in real time.

3 Results and Discussion

The system is working, the result can be monitored and displayed from the NodeMCU ESP32, apps and information data notification at the back-end smartphone device. All the information data will be set up in the Arduino IDE as it can be displayed in apps. Figure 3 shows a display of measuring the values of the NEO-6M GPS.

Fig. 2 Project flowchart

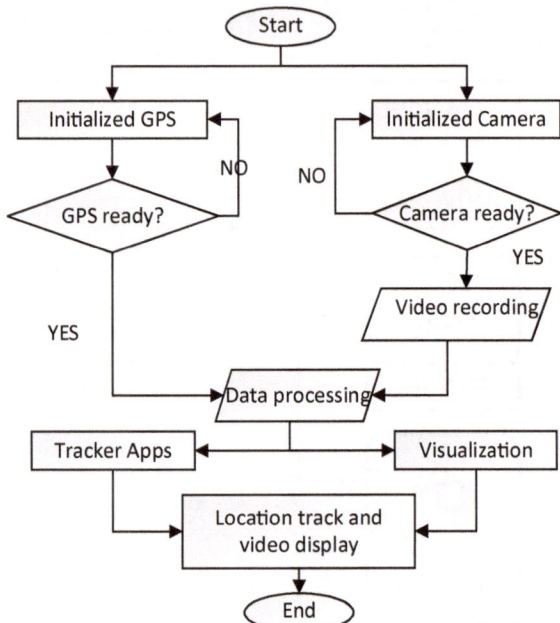

The most important part of the system is the GPS connection to obtain the value of the latitude and longitude position of the user to substantiate that the NodeMCU ESP32 is able to communicate with the sensor that is attached to it. To check the functionality of the sensor, the reading of the sensor should have appeared on the serial monitor.

Figures 4 and 5 show the position tracking and video recording captured by the Blynk app. When the microcontroller detects the location data of the surrounding, the Blynk will display a coordinate on the phone. Concurrently, the ESP32 camera will display a view of the surrounding through the Blynk app. The detected light

Fig. 3 Serial monitor from Arduino IDE for NEO_6M GPS coordinate

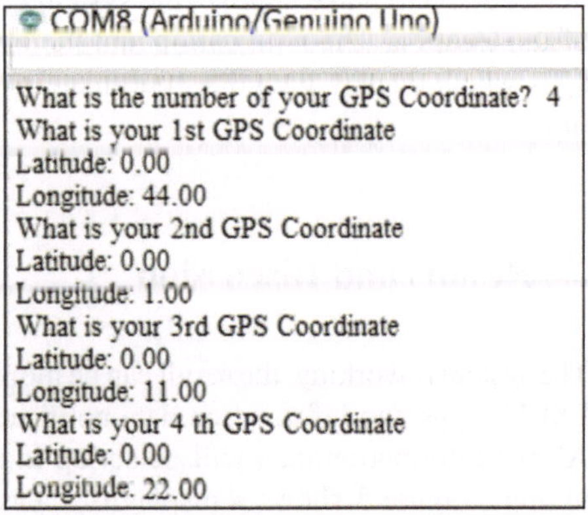

```
COM8 (Arduino/Genuino Uno)

What is the number of your GPS Coordinate? 4
What is your 1st GPS Coordinate
Latitude: 0.00
Longitude: 44.00
What is your 2nd GPS Coordinate
Latitude: 0.00
Longitude: 1.00
What is your 3rd GPS Coordinate
Latitude: 0.00
Longitude: 11.00
What is your 4 th GPS Coordinate
Latitude: 0.00
Longitude: 22.00
```

intensity parameter in terms of brightness and darkness from the ESP camera is shown in Fig. 6. When the surrounding conditions are dark, the light intensity is an average of 47 Wm^{-2}, while the average value is 373.1 Wm^{-2} when the surrounding conditions are bright.

From the GPS, the user's position is only in terms of latitude and longitude coordinates, to calculate the user's distance, the following formula is used [9, 10]:

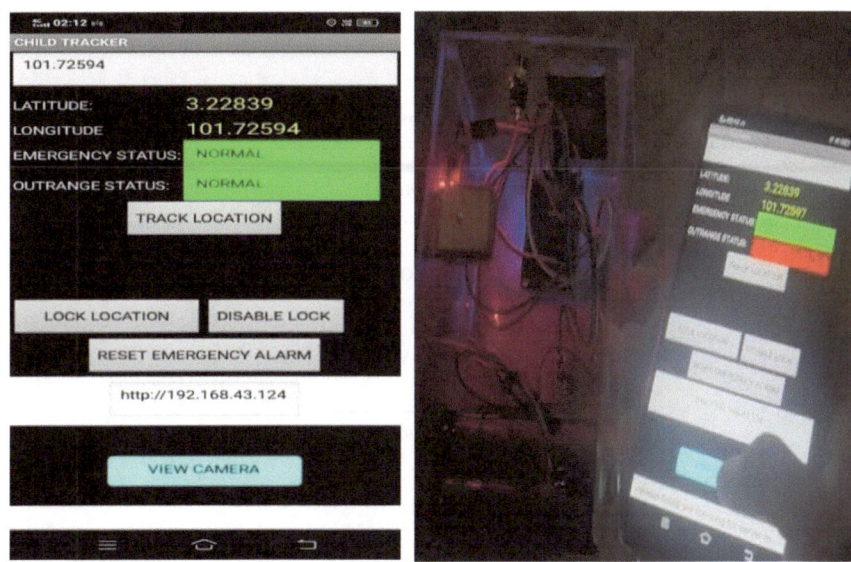

Fig. 4 Position tracking by Blynk app

Fig. 5 ESP camera video recording by Blynk app

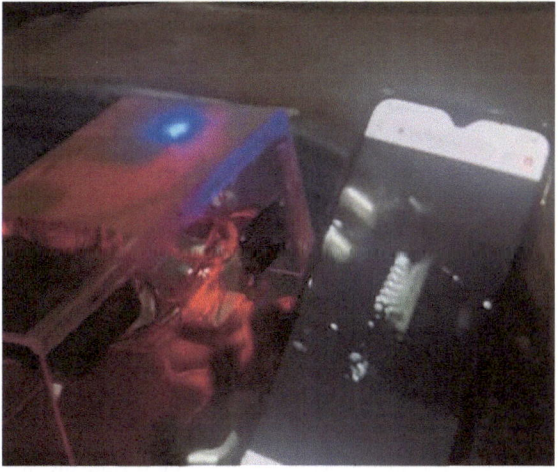

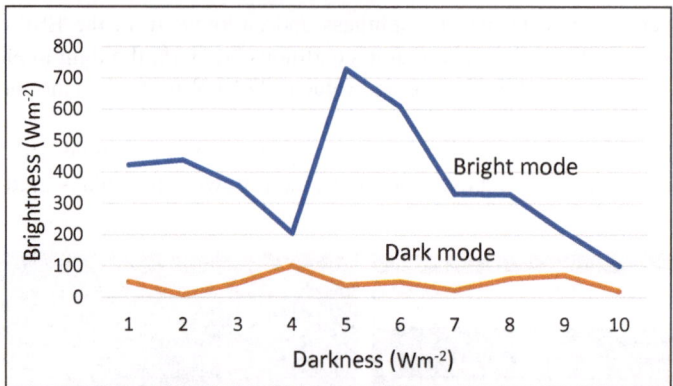

Fig. 6 Light intensity from ESP camera video recording

$$x = \Delta\lambda \cdot \cos\varphi m \tag{1}$$

$$y = \Delta\varphi \tag{2}$$

$$d = R \cdot \sqrt{\left(x^2 + y^2\right)} \tag{3}$$

where

$$\mathrm{var}\, x = (\lambda2 - \lambda1) * \mathrm{Math}\cdot\cos((\varphi1 + \varphi2)/2)$$
$$\mathrm{var}\, y = (\varphi2 - \varphi1)$$
$$\mathrm{var}\, d = \mathrm{Math}\cdot\mathrm{sqrt}(x * x + y * y) * R$$
$$\mathrm{var}\, R = 6371\, e3 \text{ in m}$$
$$\mathrm{var}\, \varphi1 = \mathrm{lat}\, 1 \text{ in Radians}$$
$$\mathrm{var}\, \varphi2 = \mathrm{lat}\, 2 \text{ in Radians}$$
$$\mathrm{var}\, \Delta\varphi = (\mathrm{lat}\, 2 - \mathrm{lat}\, 1) \text{ in Radians}$$
$$\mathrm{var}\, \Delta\lambda = (\mathrm{lon}\, 2 - \mathrm{lon}\, 1) \text{ in Radians}$$

4 Conclusion

In this paper, the development of a tracking device with an android-based GPS integration instead of using a GPS network alone is presented. Nowadays, the tracking system is useful for security and monitoring the whereabouts of family members in real time. Experimental implementation and testing with the GPS coordinate unit intensity parameter in terms of brightness and darkness from the ESP camera is shown in Fig. 6. When the surrounding conditions are dark, the light intensity is an average of 47 Wm^{-2}, while the average value is 373.1 Wm^{-2} when the surrounding conditions are bright.

From the GPS, the user's position is only in terms of latitude and longitude coordinates, to calculate the user's distance, the following formula is used [9, 10]:

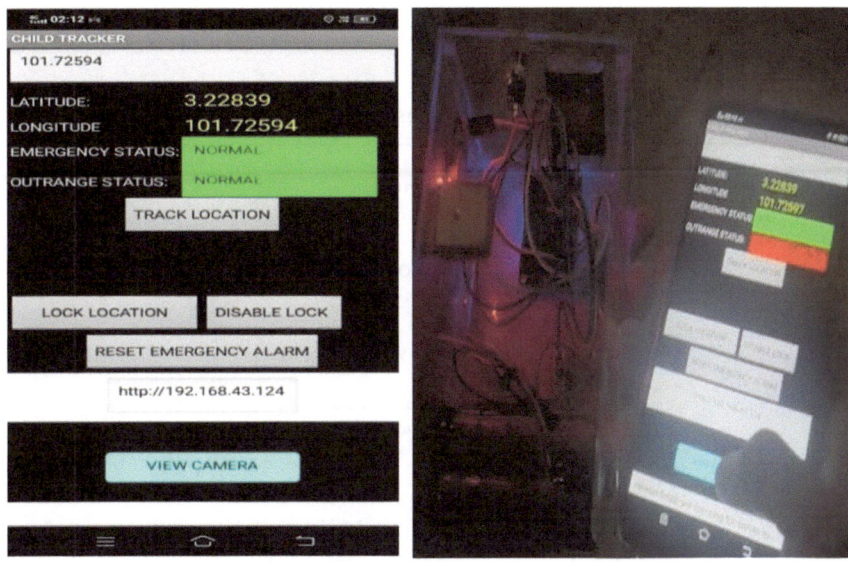

Fig. 4 Position tracking by Blynk app

Fig. 5 ESP camera video recording by Blynk app

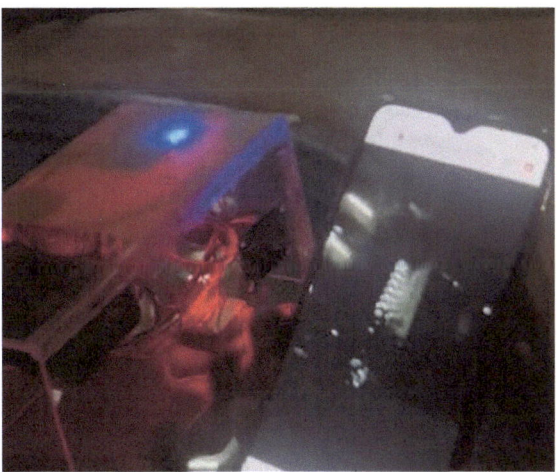

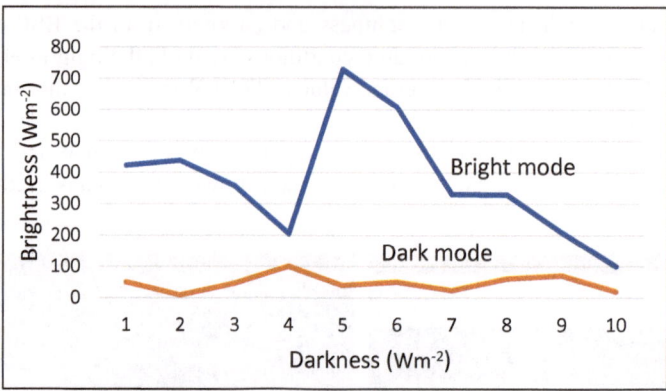

Fig. 6 Light intensity from ESP camera video recording

$$x = \Delta\lambda . \cos\varphi m \qquad (1)$$

$$y = \Delta\varphi \qquad (2)$$

$$d = R . \sqrt{\left(x^2 + y^2\right)} \qquad (3)$$

where

$$\text{var } x = (\lambda2 - \lambda1) * \text{Math . } \cos((\varphi1 + \varphi2)/2)$$
$$\text{var } y = (\varphi2 - \varphi1)$$
$$\text{var } d = \text{Math . sqrt}(x * x + y * y) * R$$
$$\text{var } R = 6371 \, e3 \text{ in m}$$
$$\text{var } \varphi1 = \text{lat 1 in Radians}$$
$$\text{var } \varphi2 = \text{lat 2 in Radians}$$
$$\text{var } \Delta\varphi = (\text{lat 2} - \text{lat 1}) \text{ in Radians}$$
$$\text{var } \Delta\lambda = (\text{lon 2} - \text{lon 1}) \text{in Radians}$$

4   Conclusion

In this paper, the development of a tracking device with an android-based GPS integration instead of using a GPS network alone is presented. Nowadays, the tracking system is useful for security and monitoring the whereabouts of family members in real time. Experimental implementation and testing with the GPS coordinate unit have been carried out several times to demonstrate the tracking operation. The project has achieved a functional location monitoring and tracking system. The GPS coordinate data can be interfaced with the Arduino mobile phone where these coordinates are mapped on a Google map. However, the limitation of this system is that the ESP32 requires a stable Wi-Fi network to transmit data from the GPS module. It may affect the accuracy of data location in real time.

Future work should be conducted to research wireless data transmission network infrastructure that can adapt to the long-range transmission distance of Wi-Fi signals generated by NodeMCU ESP32 camera and ESP32. In the future, LoRa transmission technology systems can be well implemented, allowing them to operate over longer distances of more than two kilometers at lower costs and with less power consumption.

Acknowledgements The authors wish to thank Universiti Kuala Lumpur British Malaysian Institute and Center for Research and Innovation, CoRI of Universiti Kuala Lumpur, Malaysia, for the support given to the success of this project.

References

1. Z. Khaslan, N.H.M. Yunus, M.S.M. Nadzir et al., IoT-based indoor air quality monitoring system using SAMD21 ARM cortex processor. Adv. Mater Eng. Technol. **162**, 245–253 (2022)
2. M.A. Kamarudin, N.H.M. Yunus, M.R.A. Razak et al., Development of Blynk IoT platform weather information monitoring system. Adv. Mater Eng. Technol. **162**, 295–305 (2022)
3. M.I.A. Suhaidi, N.H.M. Yunus, Development of Blynk IoT-based air quality monitoring system. J. Eng. Technol. **9**, 63–68 (2021)
4. I. Hasan, S.A. Rizvi, IQMS: IoT-based QMS framework for tracking of quarantined subjects. Int. J. Inf. Technol. **14**(5), 2255–2263 (2022)
5. P. Roy, C. Chowdhury, A survey on ubiquitous WiFi-based indoor localization system for smartphone users from implementation perspectives. CCF Trans. Pervasive Comput. Interact. **4**(3), 298–318 (2022)
6. W. Li, K. Jiang, M. Li et al., BDS and GPS side-lobe observation quality analysis and orbit determination with a GEO satellite onboard receiver. GPS Solutions **27**(1), 18 (2023)
7. S. Kumar, S.S. Rao, M. Mondal, et al., Study of the atmospheric and ionospheric phenomenon using GPS-based remote sensing technique, in *Atmospheric Remote Sensing*, pp. 261–282 (2023)
8. S. Dang, H. Huang, X. Li, About the parsing of NMEA–0183 format data streams in GPS, in *Advances in Natural Computation, Fuzzy Systems and Knowledge Discovery: Proceedings of the ICNC-FSKD 2022*, Springer International Publishing, Cham, pp. 1282–1289 (2023)
9. Z.D. Shakir, J. Zec, I. Kostanic et al., User equipment geolocation depended on long-term evolution signal-level measurements and timing advance. Int. J. Electr. Comput. Eng. **13**(2), 1560 (2023)
10. B.S. Balcha, A. Tamirat, Modeling secured user's recent-past history tracking system using location information for controlling the spread of COVID-19, pp. 1–19 (2023)

Towards an Enterprise Architecture for Healthcare System and Information Technology: State of the Art and Future Trends

Ahmad Anwar Zainuddin, Chun Kit Chung, Aiman Najmi Mat Rosani, Siti Husna Abdul Rahman, Saidatul Izyanie Kamarudin, Asmarani Ahmad Puzi, and Krishnan Subramaniam

Abstract Enterprise architecture (EA) integrates and develops organisational components for strategic planning. It helps define IT component linkages and engage paramedical workers in healthcare. It also creates the greatest healthcare enterprise architectural framework. Commonly, the senior management must grasp enterprise

A. A. Zainuddin (✉) · A. A. Puzi
Department of Computer Science, Kulliyyah of Information and Communication Technology,
International Islamic University Malaysia Gombak Campus, Gombak, Malaysia
e-mail: anwarzain@iium.edu.my

A. A. Puzi
e-mail: asmarani@iium.edu.my

C. K. Chung · K. Subramaniam
Information Technology Management, Manipal GlobalNXT University, Manipal International
Malaysia, Nilai, Malaysia
e-mail: chungkit@gmail.com

K. Subramaniam
e-mail: krishnan.subramaniam@mila.edu.my

A. N. M. Rosani
Faculty of Computer Science and Information Technology, Universiti Putra Malaysia, Seri
Kembangan, Malaysia
e-mail: aimannajmi13@gmail.com

S. H. A. Rahman
Faculty of Computing and Informatics, Multimedia University, Cyberjaya, Malaysia
e-mail: siti.husna@mmu.edu.my

S. I. Kamarudin
College of Computing, Informatics and Media, Universiti Teknologi MARA, Shah Alam,
Selangor, Malaysia
e-mail: saidatulizyanie@tmsk.uitm.edu.my

© The Author(s), under exclusive license to Springer Nature Switzerland AG 2024   49
A. Ismail et al. (eds.), *Tech Horizons*,
SpringerBriefs in Applied Sciences and Technology,
https://doi.org/10.1007/978-3-031-63326-3_7

architecture to choose a business-IT-aligned organisation. The study examines healthcare organisations' EA adoption issues before, during, and after. To fix these difficulties and streamline the process. The report recommends healthcare EA implementation. To improve implementation, these principles streamline processes, stimulate collaboration, and improve patient outcomes. EA boosts healthcare performance, but IT-human interaction must be reviewed. The healthcare businesses must overcome challenges to deploy smoothly and achieve great results.

Keywords Enterprise architecture · TOGAF · TEAF · Healthcare

1 Introduction

Healthcare is changing rapidly nowadays. Information management and digital health frameworks have changed most. Healthcare firms are more likely to use information technology (IT) in their processes [1]. Technology adoption has enhanced healthcare system EA adoption. Technology adoption and EA adoption are driven by their benefits. The organisational structure is simplified, internal and external communication is simplified and standardised, and operational costs are decreased. EA improves IT-business strategy alignment [2]. In practise, if the EA implementation is not done correctly or according to business requirements, it might raise complications and be a terrible investment. Unmanaged hazards can become problems.

According to case studies, digitised healthcare services with EA implementation provide patients with greater flexibility and superior services. One can arrange a medical appointment on his phone at his preferred time without visiting the clinic or hospital or calling them. It saves carer and care seeker time and adds convenience and flexibility [3].

EA has been implemented in healthcare systems, but few studies have examined the problems. This study examines healthcare organisation difficulties. This paper's greatest contribution is its attention on likely setbacks and hurdles healthcare systems organisations will experience before, during, and after EA adoption to help future EA implementations are simplified, streamlined, and seamless. We conducted a systematic literature review (SLR) for this purpose [4]. Moreover, preliminary work has been published in [5].

2 Theoretical Framework—Literature Review

2.1 Study Context

Despite the growing acceptance of EA in businesses [6], many have struggled with its complexity and customisation. Technology adoption is always difficult [7]. Organisations confront comparable issues, but many only recognise them when they start implementing. Understanding technology and EA implementation is crucial. This study examines how healthcare system organisations implemented EA frameworks including Zachman architecture [8], Federal Enterprise Architecture framework [9], and Open Group Architecture Framework [10]. This study excludes simple technical challenges for healthcare systems. This study also includes hospitals, NGO organisations, nursing homes, rehab clinics, community medical centres, pharmacies, postpartum care centres, and other institutions that prevent, diagnose, and treat physical and mental illnesses. Although though the most frequent EA framework for healthcare systems organisations is not the main topic of this article, we study it and consider if its inherent disadvantages will affect the issues healthcare systems organisations will encounter.

2.2 Related Works

There are significant findings in [11] describing EA implementation in the hospital scenario, which reported that IT implementation failure on such organisations occurs frequently. However, to date there are still a lack of works conducted regarding challenges of EA implementation specifically for healthcare organisations. Therefore, this study aims to provide a serious exploration on the topic of the challenges healthcare systems organisations will face regarding EA implementation as well as provide relevant new research perspectives.

3 Research Methodology

3.1 Systematic Literature Review

Since the objective of this study is to explore the challenges that organisations face in healthcare systems, we decided to achieve this through analysing key literature elements. Therefore, the systematic literature review methodology is the most appropriate research methodology [12]. SLR is especially useful in the scientific setting because we can obtain the current scientific view on topic of EA and healthcare systems via the use of secondary research. The primary steps of an SLR include the following:

- Protocol definition
- Reference search
- Screening
- Data evaluation
- Interpretation of data

3.2 Planning of Review

The selection of references for this study was based on four inclusion criteria. First, the references required to focus mostly on EA and healthcare systems; single-topic references were disregarded. The second requirement was that the publications had to have been published between 1998 and 2021. Finally, the required document type was either an article, a conference proceeding, a journal, or a book. The references must also be available in English. The research also utilised an exclusion criterion, which involves the elimination of irrelevant material. This included references with "Enterprise Architecture" or "EA" in the title that did not pertain to the study's focus on EA.

3.3 Execution of Review

Table 1 displays the classification of the research sub-questions that were addressed in this body of work. In order to make the presentation of the findings easier, we have arranged all of the problems that have been found according to [13], which is based on four different sub-questions. However, it is important to remember that not all challenges will neatly fit into one particular category, and some challenges may overlap with those of other categories. In situations like these, we have decided which category is most applicable by looking at the overarching characteristics of the problem. In order to improve the quality of our SLR list even further, we have conducted further research, which is summarised in Table 2.

3.4 Difference Enterprise Architecture Framework

To fully comprehend the challenges involved, it is imperative to have a comprehensive understanding of the EA frameworks. Various architecture frameworks have been utilised in developing an EA model for organisations, and have been studied by scholars and academics in numerous academic papers. In the following section,

Table 1 Research sub-questions

Sub-questions	Categorization
Is the challenge related to the business nature such as government restriction or EA vendors/consultants?	Environmental challenge
Is the challenge related to EA tools, infrastructure, and configuration in customizing to organizations' needs?	Technical challenge
Is the challenge related to staff resistance, communication, and training?	Organizational challenge
Is the challenge related to management and leadership, project team?	Managerial challenge

Table 2 Evolution of the number of references

Steps	Filter	Number of Articles
1	Studies retrieved from online databases	40 from Google scholar+ 22 from Scopus 2 from ScienceDirect 17 from ResearchGate
2	Studies after excluding duplication based on title	37
	Studies after excluding duplication based on content	35
3	Studies after excluding irrelevant enterprise architecture	35
4	Studies after excluding no access through Sci-hub	28
4	Studies with enterprise architecture "problems", "issues" or "challenges" specified	23

we will examine some of the prominent frameworks commonly employed in the development of an EA model.

3.4.1 Zachman Framework

It is common knowledge that Zachman developed a framework for enterprise architecture. Its formal and structured approach helps firms understand their business. A two-dimensional schema with a classified table shows the intersection of two historic classes. Primitive interrogatives—Why, How, What, Who, Where, and When—form the first dimension. Reification—turning abstract ideas into instantiations—creates the second dimension. Identification, Definition, Representation, Specification, Configuration, and Instantiation comprise this dimension. Zachman's framework is not a data collection or management method. An ontology organises various architecture artefacts such as design documents, specifications, models, regulatory documents, and more. The framework dictates who create the artefacts and what challenges they address. A UI design document for UX team members designing a health

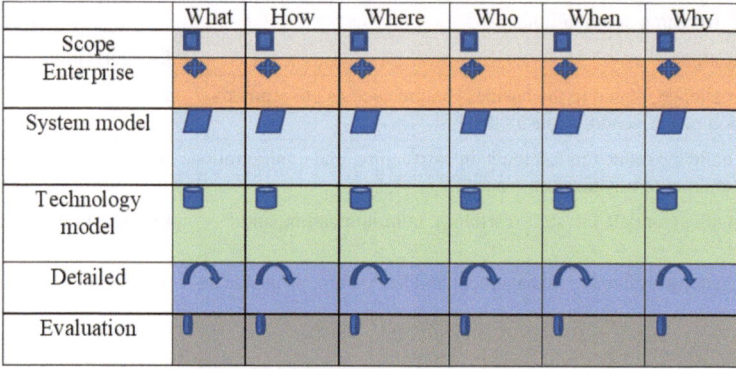

	What	How	Where	Who	When	Why
Scope	▪	▪	▪	▪	▪	▪
Enterprise	◆	◆	◆	◆	◆	◆
System model	▰	▰	▰	▰	▰	▰
Technology model	▢	▢	▢	▢	▢	▢
Detailed	↻	↻	↻	↻	↻	↻
Evaluation	▮	▮	▮	▮	▮	▮

Fig. 1 Visual of a typical Zachman framework

app for the company may address interface design. Numerous firms have adopted the Zachman framework to create a logical structure that categorises and organises business activities, providing a complete perspective of the organisation. The framework creates an organisational infrastructure that supports planning, decision-making, and communication. A typical Zachman Framework is visually represented in Fig. 1. It also develops infrastructure for the organisation that facilitates in the following:

- Designing an information system
- Developing an information system
- Integrating the information system
- Managing the information system
- Accessing the information system

3.4.2 Federal Enterprise Architecture Framework (FEAF)

Enterprise-level architecture FEAF is intended for US federal government. Despite its supposed benefits, non-federal groups employ it. FEAF promotes management, strategy, business, and technology integration. It also states that organisation design should include integration. Performance improvement should also be a priority. FEAF develops as government agencies needed more technological integration. Most government agencies have missions, and they believed that information technology could help them fulfil them faster and simpler. EAs do this. It describes the organisation's current situation and its prospects after EA implementation. The US Office of Management and Budget, E-Government, and IT developed FEAF. FEAF's main goal is a well-developed EA in US government agencies. In 1996, EA became a strategic practise for corporations, leading to the development of numerous EA frameworks. The US government wanted FEAF to standardise IT procurement and deployment across federal organisations. This would standardise federal agency IT frameworks. As one agency will depend on the other, IT implementation must be uniform to avoid misunderstanding and differentiation. This reduced federal spending and improved US citizen services. Federal agencies must follow FEAF's standards and designs.

Reference models determine FEAF's architecture. Performance, Business, Data, Application, Infrastructure, and Security Reference Models are included (SRM). These models define segment architectures. PRM integrates agency strategy, business components, and investment, for instance. This helps predict investment outcomes. DRM sets data access and storage standards. ARM sets application and system standards. This comprises system and application technologies. IRM standards cover cloud and network infrastructure (on-premises infrastructure). It defines and mandates infrastructure technology.

3.4.3 Treasury Enterprise Architecture Framework (TEAF)

TEAF is a framework built around Zachman's framework. The US Department of the Treasury founded it in 2000 to standardise IT policies and initiatives inside the government. TEAF is exclusively developed for enterprises with treasury departments and only supports business procedures inside those departments. TEAF has two primary goals: to develop business processes that comply with regulatory standards while implementing technology, and to adapt to continually evolving technologies without affecting legislation. It is essential that the adoption and implementation of IT adhere to regulatory and statutory criteria. Data storage and privacy are excellent examples of legal and statutory criteria that IT integration must meet. TEAF provides unification and standardisation of common principles, concepts, and technology for the various treasury agencies involved in the creation of information systems. In addition, it supplies departments with a ready-made template for embracing information technology. Functional, informational, and organisational architectural views define a model for business processes, functions, and operations.

3.4.4 The Open Group Architecture Framework (TOGAF)

The Open Group Architecture Framework (TOGAF) is one of the most extensively adopted corporate architecture frameworks in the world. It was first created in 1995 to manage information systems and is based on the Technical Architect Framework of the United States Department of Defence. TOGAF's universal applicability to any sector or market area is one of its chief features. It also permits modification based on an organisation's own requirements. TOGAF's modular design enables the adoption and implementation of its components in stages. In addition, it features a well-defined content framework that helps prevent inconsistencies in architectural development outputs. TOGAF additionally provides enhanced concepts and standards to facilitate the creation of integrated architectures by multiple teams inside an organisation. It can be utilised as a stand-alone architecture or in conjunction with another style, such as TEAF, for a particular department or purpose. Architectural Development Method (ADM) is a cyclical and methodical process for designing

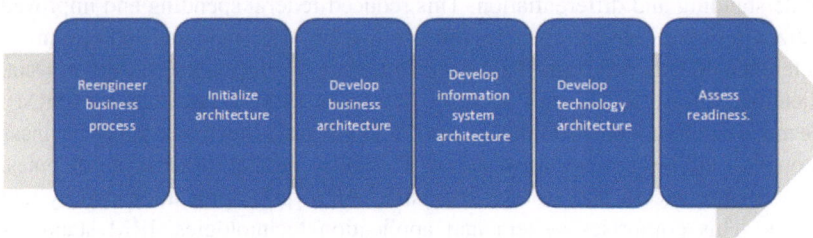

Fig. 2 Proposed model for hospital information system

corporate architecture that takes both business and IT requirements into account. The extent of modification is determined by the organisation's business requirements and the foundation architecture. TOGAF's current version is 9.2.

3.5 Role of EA in Healthcare

Electronic medical records (EMR) have prompted concerns regarding record discrepancies, medical errors, and interoperability due to a lack of standardisation, which can harm patient care. EA can provide an integrated healthcare information system (HIS) with accurate, trustworthy, and timely healthcare information for medical practitioners to make prompt and appropriate decisions. Due to unintegrated systems, healthcare technology is underutilised. Eliminate unnecessary processes to increase efficiency, simplify each business process to reduce complexity, integrate previously incompatible and heterogeneous processes into one interoperable process, and automate to reduce human intervention and error in the existing process. Figure 2 shows the suggested model's phase focus, referenced framework technique, and important steps.

4 Summary of Findings

Healthcare organisations face several concerns and challenges. EA deployment in healthcare systems organisations faces four main challenges:

1   Industrial or inter-organisational health environments.

Organisations cannot use a predefined architecture or model [14]. Healthcare systems companies vary in maturity, size, location, speciality, and technological and business modifications. EA can help provide a holistic view in the complex healthcare business [15], but it is not a panacea.

2. Data integration and access issues.

Inadequate managerial commitment, knowledge, and support. Even while management recognises EA's benefits [16] and understands its models, methodologies, principles, and frameworks, their comprehension of EA is not enough to help the organisation transform so drastically. Commitment is usually moderate.

3. Lack of clear definition of the organisation's objectives:

   (i) Environmental challenge
   (ii) Technical challenge
   (iii) Organisational challenge
   (iv) Managerial challenge.

5 Conclusion

One of the limitations for this study is the selection criteria of "healthcare". As we wanted to keep the sample to a manageable size, we narrowed the keywords to healthcare for titles only. Other keywords such as "public sector", "hospital", and "clinics" were not selected. The reviewer's perspective on identifying and categorising a problem, issue, or challenge in the references is an unavoidable limitation. These are based on our reviewer's judgement which will somehow affect the objectivity of the study. Many organisations are prepared themselves with the challenges they anticipated. It is worth noting that the real challenges are those that come as unexpected such as executing how EA communicates to external organisations.

References

1. M.W. Wichmann, An exploration of enterprise architecture research in hospitals, in *Business Information Systems Workshops*, ed. by M.W. Johannes Wichmann (Springer, Cham, 2019), pp. 89–10
2. T. Tamm, How does enterprise architecture add value to organisations. Commun. Assoc. Inf. Syst. **28**, 141–168, (2011)
3. X. Zhang, P.Y.: Patients' adoption of the e-appointment scheduling service: a case study in primary healthcare. Stud. Health Technol. Inf. **204**, 176–181 (2014)
4. B. Kitchenham, C.S., Guidelines for performing systematic literature reviews in software engineering. Technical Report EBSE 2007-001 (2007)
5. M. Rozario, A.A. Zainuddin, P. Manokaran, Analysis of a survey on the role of enterprise architecture in the field of healthcare technology by knowledge-based research. Int. J. Comput. Digital Syst. **13**(1), 547–557 (2023)
6. H. Thimbleby, Technology and the future of healthcare. J. Public Health Res. **2**(3), e28 (2013)
7. F. Armour, *Enterprise Architecture: Challenges and Implementations*, p. 217 (2007)
8. J.A. Zachman, *The Concise Definition of the Zachman Framework* (2018). https://www.zachman.com/about-the-zachman-framework
9. OMB.: FEA Consolidated Reference Model Document Version 2.3 (2007). https://web.archive.org/web/20090202182509/, http://www.whitehouse.gov/omb/as

10. N. Dedić, FEAMI: a methodology to include and to integrate enterprise architecture processes into existing organizational processes. IEEE Eng. Manag. Rev. **48**(4), 46–54 (2020)
11. J.S. Faizal Pasaribu, Designing enterprise architecture in hospitals group, in *International Conference on Information and Communications Technology (ICOIACT)* (2019)
12. B.A. Kitchenham, O.P. Brereton, D. Budgen, Protocol for extending an existing tertiary study of systematic literature reviews in software engineering (2017)
13. K.S. Negin Banaeianjahromi, Understanding obstacles in enterprise architecture development, in *European Conference on Information Systems (ECIS)*, Istanbul, Turkey (2016)
14. S.A. Cresswell Kathrin M.: Health information technology in hospitals: current issues and future trends. Future Hosp. J. **2**(1), 50–56 (2015)
15. H. Plessius et al., Towards an enterprise architecture benefits measurement instrument. International Conference on Advanced Information Systems Engineering, Lecture Notes in Business Information Processing **215**, 363–374 (2015)
16. I. Abunadi, Enterprise architecture best practices in large corporations. Information **10**(10), 293 (2019)

A Regression Analysis for Predicting Student Academic Performance

Zuraini Zainol, Puteri Nor Ellyza Nohuddin, Husna Sarirah Husin, Ummul Fahri Abdul Rauf, and Muhammad Yazid Abdul Mutalib

Abstract The aim of the study is to identify the factors that accurately predict academic performance and the contribution that each factor makes to overall academic success. The collected dataset consists of 21 attributes for 97 students in one of the public universities in Malaysia. Several data pre-processing and feature selection tasks have been performed to ensure the quality of the data. A regression model is developed to predict the cumulative grade point average (CGPA) using the selected variables. The result discovered that variables such as gender, absence rate and GPA affect the CGPA of the students. Model evaluation also proves that it can be utilized to predict the CGPA of students. This study is expected to help educational institutions, particularly academic advisors, in identifying students who are at risk of failure. Thus, an early effective program can assist students in improving their academic performance.

Keywords Academic performance · Regression analysis · Educational data mining · Prediction · Multiple linear regression

Z. Zainol (✉) · M. Y. A. Mutalib
Computer Science Department, Universiti Pertahanan Nasional Malaysia, Kuala Lumpur, Malaysia
e-mail: zuraini@upnm.edu.my

M. Y. A. Mutalib
e-mail: muhammadyazid250@gmail.com

P. N. E. Nohuddin
Faculty of Business, Higher College of Technology, Sharjah, UAE
e-mail: pnohuddin@hct.ac.ae

H. S. Husin (✉)
Universiti Kuala Lumpur Malaysian Institute of Information Technology, Kuala Lumpur, Malaysia
e-mail: sarirah@unikl.edu.my

U. F. A. Rauf
Pusat Asasi Pertahanan, Universiti Pertahanan Nasional Malaysia, Kuala Lumpur, Malaysia
e-mail: ummul@upnm.edu.my

© The Author(s), under exclusive license to Springer Nature Switzerland AG 2024
A. Ismail et al. (eds.), *Tech Horizons*,
SpringerBriefs in Applied Sciences and Technology,
https://doi.org/10.1007/978-3-031-63326-3_8

1 Introduction

The field of education is always looking for new methods for improving students' grades. For educators, politicians and researchers to create successful solutions to improve students' learning outcomes, they must first have a thorough understanding of the components that contribute to academic performance. Students' demographics, social background and prior academic success are only a few of the many aspects that have been investigated in relation to their performance in school. New evidence, however, suggests that other characteristics may also be important in predicting academic performance.

Regression analysis is one method for predicting academic performance. Regression analysis is to investigate the relationship between one or more predictor variables and a dependent variable (DV). In the case of student academic performance, the DV is usually a measure of achievement, such as the cumulative grade point average (CGPA), while the predictor variables can be anything from demographic characteristics to past academic performance. The aim of this research is to investigate, by use of regression analysis, the connection that may be established between a number of different predictor variables and the academic performance of students. To be more specific, it is to determine which variables are the most accurate predictors of academic performance and the degree to which each variable contributes to total academic success.

To accomplish the aim of this study, a dataset is compiled from a selection of students who are currently enrolled at Universiti Pertahanan Nasional Malaysia (UPNM). The dataset contains information on students' demographic characteristics (such as age, gender and race/ethnicity), socioeconomic status (such as parental income and education), previous academic performance and other relevant variables. The research consists of a few data and analysis processes using the multiple linear regression (MLR) model. First, the data is processed as descriptive statistics to summarize the characteristics of the data sample and the distribution of the variables. Then, the study investigates a correlation analysis to determine the bivariate associations that exist between the predictor variables and the variable that is being studied (the dependent variable).

The findings of this study contribute to the existing body of knowledge on student academic performance and may have ramifications for educators and policymakers in the real world. This study has the potential to inform the creation of interventions and policies targeted at enhancing student accomplishment since it will uncover the factors that are the best predictors of academic success. In addition, the results of this study could provide a basis for further investigation in the form of research in the future that investigates more complicated models and the interactions between predictor variables.

2   Literature Review

There has been a rapid increase in identifying the students' performance, particularly using data mining (DM) techniques, typically known as educational data mining (EDM). The ability to extract new knowledge from a large amount of students' data has made EDM ubiquitous and widely used. Generally, EDM uses DM techniques such as clustering [1–4], classification [5–7] and association rules [8] to retrieve data in educational systems. In terms of predicting student academic performance, many studies have been performed. Researchers used different factors and attributes of the students to predict their performance and learning outcomes. Research to improve the prediction accuracy of student academic performance employed are naïve Bayes, logistic regression, k-nearest neighbor and random forest and were used to generate predictive models [9]. Another study proposes an ensemble meta-based tree model (EMT) classifier technique for predicting the student performance and their experimental results show that the EMT as the ensemble technique gained a high accuracy performance reaching of 98.5% (or 0.985) [10]. In another study to predict student grades, the results show that k-NN achieves the highest accuracy with 77.36%, where both feature selection and parameter optimization are applied [11].

Besides identifying the factors and attributes of the students, [12] examined the student's assignment submission behavior and students' behavioral patterns before a homework due date to predict their academic performance. They found that linear support vector machine is the best classifier among others in terms of continuous features, and neural network in categorical features, where categorical features tend to perform slightly better than continuous. On the other hand, there have been studies that take into account of students' performance, motivation and resilience to identify students who are more likely to experience academic failure [13].

3   Framework of Student's Academic Performance Prediction

Figure 1 illustrates the framework of student's academic performance prediction (SAPP) model for predicting the student's academic performance. It consists of three main modules: (i) data collection, (ii) data pre-processing and feature selection and (iii) modeling and predicting CGPA using MLR. The detailed explanation of each module will be discussed in Sect. 4.

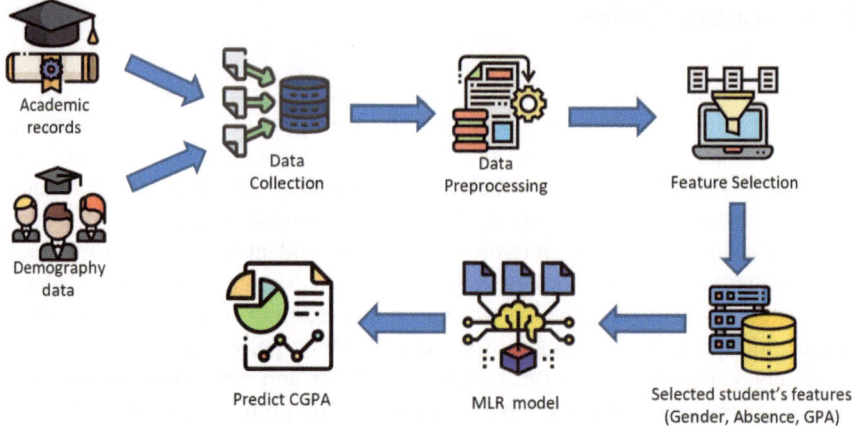

Fig. 1 Framework of the SAPP model

4 Experimental Setup and Analysis

4.1 Data Collection

There are two different sets of data used in this study. The first dataset was retrieved from the Academic Management Division, UPNM. It contained student's academic data such as grades of courses for the fifth semester of Computer Science students in the academic year 2022–2023. The second dataset related to demographic data which was collected using the questionnaires through the application of Google forms. The collection of questionnaires was created based on the study of [14–16]. They examined numerous factors (attributes) used for predicting the student academic performance such as CGPA, grade point average (GPA), demographics information (e.g., age, gender, parents' status, mother's and father's education, student's activities).

4.2 Data Pre-processing and Feature Selection

The second module focused on pre-processing the collected data. This task is crucial as it produces high-quality data [17] and error-free analysis [18]. Three main data pre-processing such as data filtering, label encoding and feature selection were applied on the dataset. From the 21 attributes, only 20 attributes are left (after removing the attribute 'Matric'). Table 1 shows the lists of the attributes with its description. In supervised learning, most of machine learning algorithms performed better on the numerical data. Therefore, the 18 categorical variables (attributes) such as 'Gender', 'Race, 'TypeStudent', 'NumSiblings', 'ParentAlive'. 'ParentStatus', 'FQualifica-tion', 'MQualification', 'FJobStatus', MJobStatus', 'FamilyStatus', 'Scholarship',

'SAccomodation', 'Activities', 'AssignmentTime', 'SocmedTime', 'Absence' and 'Part time' must be encoded before they can be applied in the model. Label encoding is one of pre-processing techniques used for handling the categorical data. In this study, all the categorical string value were converted into integer value using some Python code. The next step is to select a set of the independent variables (IV) or features that are most relevant to the DV. A large dataset with numerous features can cause time-consuming for the model to execute. In this experiment, CGPA was selected as a DV. Correlation is one of the most popular feature selection techniques used for numerical features and a numerical target attribute. It is often used to measure how two attributes change together. In this experiment, attributes with a correlation coefficient value above 0.3 were set up using the Python command line. Three features such as 'GPA' (0.883043), 'Absence' (0.456373) and 'Gender' (0.335041) were identified as highly correlated with the DV (CGPA). Therefore, these features were applied in the development of MLR model.

Table 1 List of attributes and description

No.	Attributes	Values	Data type
1	Gender	Male, female	Categorical
2	Race	Malay, Chinese, Indian	Categorical
3	TypeStudent	Cadet, Civilian, PALAPES	Categorical
4	NumSiblings	1, between 2–5, more than 5	Categorical
5	ParentsAlive	Yes, no	Categorical
6	ParentsStatus	Married, divorce, widowed	Categorical
7	FQualification	Graduate, secondary, elementary, no education	Categorical
8	MQualification	Graduate, secondary education, elementary, no education	Categorical
9	FJobStatus	Currently on service, housewife, retired, none	Categorical
10	MJobStatus	Currently on service, retired, none	Categorical
11	FamilyStatus	B40, M40, T20	Categorical
12	Scholarship	Yes, No	Categorical
13	SAccomodation	In campus, rent a room, stay with family	Categorical
14	Activities	Sports, student representation council (MPP), clubs	Categorical
15	AssignmentTime	Less than 1 h, between 2–4 h, more than 4 h	Categorical
16	SocmedTime	Less than 1 h, between 2–4 h, more than 4 h	Categorical
17	Absence	Less than 3 days, between 4 7 day, more than 7 days	Categorical
18	Part time	Yes, no	Categorical
19	GPA	Grade point average (from 0 to 4.00)	Numerical
20	CGPA	Cumulative grade point average (from 0 to 4.00)	Numerical

4.3 Modeling and Predicting CGPA Using MLR

The input of the MLR model were 'Gender', 'Absence' and 'GPA' was assigned to a variable 'x'. The 'CGPA' was assigned to variable 'y'. The dataset was split into 60:40 ratio. 60% of the dataset was used for training the model, and 40% was used for the testing purpose. Based on the experiment, the formula for the MLR model is presented in Table 2. The MLR produced the intercept as -0.2343 and the slopes are -0.0248 (Gender), -0.0421 (Absence) and 1.0677 (GPA). Based on Table 2, all variables have negative correlation except GPA. The model produced an accuracy of 84.79%. Based on the result (Table 3), the model shows positive correlation of 0.8479, and minimal error of RMSE (0.08627) and MAE (0.06335). This indicates the model explains almost 85% variability of the data. Figure 2 shows a graph of the predicted CGPA against the actual CGPA. For example, the actual CCPA is 3.12 and predicted is 3.16 (see the value in highlighted yellow).

Table 2 MLR model

	Independent variables	Equation for the MLR model
1	Gender	$-0.2343 - 0.0248 \times$ Gender $- 0.0421 \times$ Absence $+ 1.0677 \times$ GPA
2	Absence	
3	GPA	

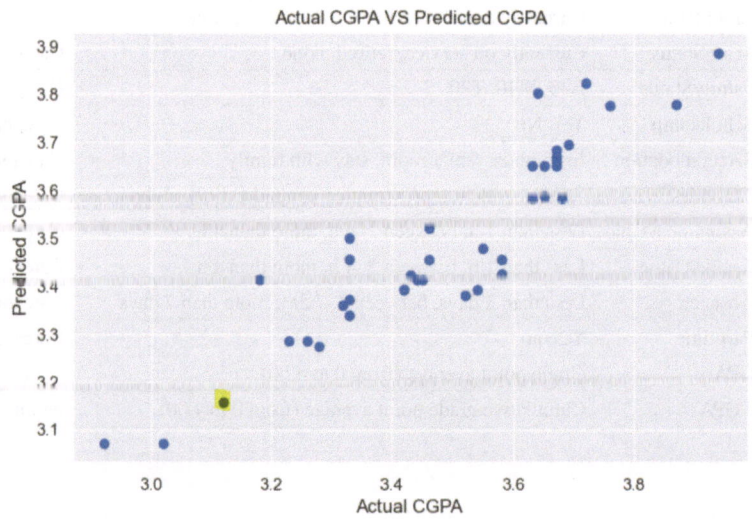

Fig. 2 Actual CGPA versus predicted CGPA

Table 3 Summary of performance result

Model	RMSE	MAE	R-squared
MLR model	0.08627	0.06335	0.8479

5 Conclusion

This study proposed a regression model for predicting student's performance using the MLR algorithms. The dataset used in this study comprises 97 undergraduate students with 21 attributes from UPNM's Computer Science students. The experimental results show that attributes such as GPA, absence and gender were identified as the most important variables in predicting students' academic performance. The MLR model shows a positive correlation of 0.8479, and minimal error of RMSE (0.08627) and MAE (0.06335). This indicates the model explains almost 85% variability of the data. The future work will highlight on expanding the dataset by adding more variables (features) related to student academic performance and further improving regression performance by applying appropriate feature selection techniques.

Acknowledgements We are grateful to Malaysian Institute of Information Technology for providing financial support for this research and SF0139-UPNM/2023/SF/ICT/4 for supporting this research.

References

1. S. Fida, N. Masood, N. Tariq, F. Qayyum, A novel hybrid ensemble clustering technique for student performance prediction. J. Univers. Comput. Sci. **28**(8), 777–798 (2022)
2. P.N.E. Nohuddin, Z. Zainol, A. Nordin, Monitoring students performance using self organizing map trend clustering. Int. J. Def. Sci. Eng. Technol. **1**(1), 50–56 (2018)
3. O. Iatrellis, I.K Savvas, P. Fitsilis, V.C. Gerogiannis, A two-phase machine learning approach for predicting student outcomes. Educ. Inf. Technol. **26**, 69–88 (2021)
4. A.M. Abdo, N.M.A. Rasid, N.A.H.M. Badli, S.N.A. Sulaiman, S. Wani, Z. Zainol, Student's performance based on e-learning platform behaviour using clustering techniques. Int. J. Perceptive Cogn. Comput. **7**(1), 72–78 (2021)
5. S.B. Rahayu, N.D. Kamarudin, Z. Zainol, Case study of UPNM students performance classification algorithms. J. Eng. Technol. **7**(4.31), 285–289 (2018)
6. Y.S. Alsalman, N.K.A. Halemah, E.S. AlNagi, W. Salameh, Using decision tree and artificial neural network to predict students academic performance, in *Paper presented at the 10th International Conference on Information and Communication System* (2019), pp. 104–109
7. F. Jauhari, A.A. Supianto, Building student's performance decision tree classifier using boosting algorithm. Indonesia J. Electr. Eng. Comput. Sci. **14**(3), 1298–1304 (2019)
8. P.N.E. Nohuddin, Z. Zainol, M.H.A. Hijazi, Study of B40 schoolchildren lifestyles and academic performance using association rule mining. Ann. Emerg. Technol. Comput. **5**(5), 60–68 (2020)
9. G. Ramaswami, T. Susnjak, A. Mathrani, J. Lim, P. Garcia, Using educational data mining techniques to increase the prediction accuracy of student academic performance. Inf. Learn. Sci. **120**(7/8), 451–467 (2019)

10. A. Almasri, E. Celebi, R.S. Alkhawaldeh, EMT: ensemble meta-based tree model for predicting student performance. Sci. Program (2019)
11. W.W. Damopolii, N. Priyasadie, A. Zahra, Educational data mining in predicting student final grades. Int. J. **10**(1) (2021)
12. D. Hooshyar, M. Pedaste, Y. Yang, Mining educational data to predict students' performance through procrastination behavior. Entropy **22**(1), 12 (2019)
13. A. Sarra, L. Fontanella, S. Di Zio, Identifying students at risk of academic failure within the educational data mining framework. Soc. Ind. Res. **146**, 41–60 (2019)
14. O. El Aissaoui, Y. El Alami El Madani, L. Oughdir, A. Dakkak, Y. El Allioui, A multiple linear regression-based approach to predict student performance, in *Paper presented at Advanced Intelligent System Sustainable Development*, vol. 1102 (2020), pp. 9–23
15. K.I.M. Fadilah, Z. Zainol, M. Ebrahim, A.S.H. Lee, Covid-19 effect on undergraduate computing students' performance at higher education: pilot study. *Paper presented at the 6th IEEE International Conference on Recent Advanced and Innovations in Engineering*, pp. 1–6 (2021)
16. E. Alyahyan, D. Düştegör, Predicting academic success in higher education: literature review and best practices. Int. J. Educ. Technol. High. Educ. **17**(3), 1–21 (2020)
17. S.A. Alwarthan, N. Aslam, I.U. Khan, Predicting student academic performance at higher education using data mining: a systematic review. Appl. Computat. Intell. Soft Comput. (2022)
18. A. Navlani, A. Fandango, I. Idris, *Python Data Analysis: Perform Data Collection, Data Processing, Wrangling, Visualization and Model Building using Python* (Packt Publishing Ltd, 2021)

Bumiputera-Owned Small and Medium Enterprise Family Business Succession Plan: A Review

Siti Noor Kamariah Yaakop, Nooraini Othman, and Wardiah Mohd Dahalan

Abstract Transgenerational entrepreneurship entails the proactive development of a self-sustaining family-owned business that can be passed down through multiple generations, whereas family business succession refers to the transfer of business ownership and management to the next generation through inheritance. Choosing a successor in small and medium-sized enterprises requires a more comprehensive approach than in larger corporate firms due to fewer potential successors. The purpose of this research is to identify the challenges and factors that incumbents must consider when selecting a successor and implementing a succession plan. The study identifies challenges such as the involvement of an informal family member in the selection process, internal conflicts between the family and the business, the incumbent's attachment to their leadership legacy, and the successor's readiness to assume responsibility. When it comes to succession planning, education, competence, demographic factors, relationships with family members and incumbents, experience, integrity, birth order, and primogeniture are all important factors to consider.

Keywords Family-owned business · Succession plan · Succession challenges · Succession factors · Successor · Incumbent

S. N. K. Yaakop (✉)
Teknoputra Section, Universiti Kuala Lumpur Malaysian Institute of Marine Engineering Technology, 32200 Lumut, Perak, Malaysia
e-mail: sitikamariah@unikl.edu.my

N. Othman
Perdana Center, Razak Faculty of Technology and Informatics, Universiti Technologi Malaysia, Jalan Sultan Yahya Petra, Kampung Datuk Keramat, 54100 Kuala Lumpur, Malaysia
e-mail: p-nooraini@utm.my

W. M. Dahalan
Marine Electrical and Engineering Section, Universiti Kuala Lumpur Malaysian Institute of Marine Engineering. Technology Lumut, Perak, Malaysia
e-mail: wardiah@unikl.edu.my

© The Author(s), under exclusive license to Springer Nature Switzerland AG 2024
A. Ismail et al. (eds.), *Tech Horizons*,
SpringerBriefs in Applied Sciences and Technology,
https://doi.org/10.1007/978-3-031-63326-3_9

1 Introduction

Due to the Movement Control Order (MCO) in March 2020 caused by Covid-19, many government and private premises were closed globally, leading to Malaysia's economy shrinking by 17.1% in Q2 2020 [1]. This has forced households and industry players to adapt to new norms, including family businesses that contribute to the global economy [2]. As the future of the family business depends entirely on the leader, succession planning has become critical to business sustainability, but it is a complicated responsibility [3]. Despite many studies, the implementation of succession plans still faces failure in family businesses [4–6]. To ensure the ability of the business to survive, family business leaders need to lead the business more efficiently and ethically by evaluating business priorities [7], transforming into digital businesses [8], and reorganizing family business successors. This study identifies factors to consider in choosing potential family business successors.

2 Analysis of the Previous Works

2.1 Family Business

For decades, family businesses have been recognized as the oldest and distinct commercial entity [9–11]. Some of the most prosperous companies have established a dominant global presence, starting from a family-owned and operated business [12]. These businesses have a shared vision, goals, and strategies to enhance their family's financial growth while preserving family relationships. Their dedication to preserving and expanding the business allows it to be passed down from one generation to the next [13, 14]. However, various inquiries remain regarding the understanding of family businesses, including defining the term, establishing criteria, addressing challenges, and implementing succession plans to secure the family enterprise's continuity.

2.2 Family Business Criteria

When family members are involved in a business, it is often considered distinct, but it is challenging to establish a clear definition and criteria for what constitutes a family business. Prior research has presented multiple viewpoints on the definition of family business. Scholars have offered diverse interpretations of the concept in their respective studies, with some defining it based on factors like company size, ownership, and structure.

A family business is one that is majority owned, controlled, and administered by blood or marriage relatives, or that has at least 50% management and administration by blood or marriage relatives [15]. Two or more individuals from the same family members who are actively participating in the firm have been given the majority power to own, control, and then make decisions involving management, finance, and the future of their business. In fact, when family members own a majority stake in a company, they have complete control over the company's journey [16].

Balancing the ownership of a legacy business with accountability for its performance creates a complex situation for business leaders, who must manage both the family system and the business system [17, 18]. The two systems are interconnected, which further complicates matters for SME family businesses [19, 20]. The family system, characterized by emotional relationships, plays a significant role in the administration of such businesses. Additionally, family businesses tend to be smaller in size, even in Malaysia where the Chinese community owns a majority of these businesses [21, 22]. Scholars have previously defined a family business as one where the administration and operation of the firm is driven by a business vision predominantly owned by a combination of family members or a small portion thereof [23, 24].

2.3 Small and Medium Enterprise

The National Entrepreneurship Policy Report 2030 states that small and medium-sized businesses (SMEs) make up more than 90% of all businesses. They are also responsible for generating the majority of job opportunities and contributing up to 50% of a country's total value generated to GDP [25]. The definition of SMEs varies depending on the country and industry, but most countries use business size, annual sales, and the number of full-time employees to define them. In Malaysia, SMEs are defined by their sales turnover and the number of full-time employees and must be registered with the appropriate government body (https://www.smeinfo.com.my/off icial-definition-of-sme/). However, certain entities such as publicly listed companies and subsidiaries of multinational corporations are not considered SMEs.

2.4 Succession Planning

Transgenerational entrepreneurship entails passing down a family business's autonomy from generation to generation [26, 27]. Regardless of the challenges, succession planning is critical for the success, growth, and legacy of family businesses [19, 28]. Business leaders must implement a comprehensive plan to systematically and effectively transfer leadership channels, ensuring that competent successors lead the firm in the future [29, 30]. The successful implementation of a succession plan demonstrates the ability to manage and sustain transgenerational entrepreneurship. This necessitates comprehensive planning to ensure the continuity and sustainability of family businesses, as well as providing opportunities for family members to advance in their careers and identifying new leadership competencies that will improve firm performance.

Succession planning entails identifying a successor to lead the company after the incumbent retires, resigns, or dies [15, 31]. It is critical for the continuity and sustainability of family businesses, ensuring that profits continue to rise and planned employee retention and involvement is maintained [32, 33]. The goal of succession planning is to select a competent leader who can elevate the family's legacy to a higher level rather than simply deciding who will lead the company [34, 35]. Transferring capital ownership to the business's extended family is also part of the succession process, resulting in the effective transfer of family firm heritage.

To ensure the plan's success and effectiveness, all stakeholders, including the founders, family members, business leaders, shareholders, and the business environment, must commit to it and understand its significance [36, 37]. The agreement of all family members and stakeholders in the ownership of family business shares becomes a top priority on the agenda in order to achieve the goal of a time-bound plan that respects the rights of all parties involved.

2.4.1 Succession Planning Process

Succession in family businesses refers to the transfer of ownership and leadership to the next generation [4, 38]. Previous research has highlighted the significance of succession planning for long-term leadership investment, involving multiple parties, such as the incumbent, potential successors, and family members. Effective implementation requires resolving conflicts and allocating sufficient time to avoid complications in decision-making [27, 37]. However, family businesses often struggle to implement succession plans due to a lack of standardized strategies [39].

Family businesses use various succession processes, such as continuing the incumbent's leadership style, receiving advisory services from mentors or consultants, family communication and guidance, and influential individuals in the family [17, 40]. Scholars suggest that family businesses should develop protocols to guide employment, ownership transfer, and administration among family members [41, 42]. Starting succession planning early and establishing a harmonious relationship between business leaders and successors can facilitate knowledge transfer and ensure future success. Additionally, the successor's attitude and motivations toward maintaining business sustainability and family relationships also play a crucial role [43].

2.4.2 Successor Selection Factors

Choosing a successor and passing on power and inheritance to the next generation is a crucial stage in the life cycle of a family business, as noted by various researchers [44, 45]. It requires strategic planning and thorough research into the implementation strategy, criteria, and compensation for potential successors, as incumbents must identify competent and credible individuals who can effectively lead their family's legacy business.

Prior studies have examined various factors in selecting potential successors who are qualified to take over the family business, including the traditional concept of primogeniture, which favors the eldest son [44, 46]. However, other experts suggest considering birth order as well [47, 48]. Education is also a critical trait to consider, as potential successors should have formal education obtained through high school or university studies [4, 49–51]. Additionally, combining knowledge and experience is crucial to ensure that successors have credibility and the ability to run the organization effectively. The exposure gained from working with external businesses can provide valuable experience and contacts for family-owned businesses [52]. Competence and skill capabilities possessed by the potential successor gained through education and external experience can enhance decision-making abilities [36, 44, 53, 54].

The level of interaction between possible successors and incumbents/family members is also important in future successor selection. Family connections and values are key aspects of succession planning, according to research by Abdullah et al. [55] [53]. Effective communication between successors, company executives, and family members is critical in resolving conflicts in family enterprises [53, 56].

3   Research Methodology

The methodology of the study employs a qualitative approach that is consistent with the research questions and objectives of phenomenological exploration. Qualitative research methods allow for a comprehensive explanation of events as they occur, without the researcher intervening. The holistic explanation necessitates a historical examination of past events that may aid in explaining current and future trends and effects [57]. Phenomenology will be used to investigate the relationship between a person's life experience and a phenomenon, with the goal of understanding and describing the participants' interactions with everyday life practices [58].

In order to investigate the importance of having a succession plan, interviews were conducted with incumbents based on their experiences. The study's emphasis is on passing on a business to the next generation, and business owners and successors shared their experiences with succession process practices to ensure business continuity across generations. The study looks into the factors that are taken into account when selecting a successor in the process of implementing a family business succession plan.

4 Analysis of the Result

In addition, the potential successor's personal characteristics and leadership qualities are also crucial factors to consider. Previous studies suggest that a successor's
personal characteristics such as integrity, passion, adaptability, and risk-taking
propensity are critical in determining their ability to lead and manage the family
business [5, 49–51]. Leadership qualities such as vision, strategic thinking, and
decision-making abilities are also necessary for effective succession planning
[36, 44, 53, 54].

Moreover, it is essential to consider the potential successor's willingness to take
on the responsibility of running the family business. Succession planning is a long-
term process that requires commitment, dedication, and a sense of obligation to the
family legacy. Therefore, the potential successor's motivation and interest in the
family business should be evaluated to ensure that they have a genuine desire to
continue the legacy [29, 47, 48].

To summarize, selecting a qualified and capable successor for a family-owned
business necessitates a comprehensive approach that takes into account a variety of
factors such as birth order, education, experience, personal characteristics, leadership
qualities, and willingness to assume responsibility for the family business. Communication and family connections are also important in ensuring a smooth succession
process.

5 Conclusion

The vision, leadership style, communication skills, adaptability, and willingness to
learn and take risks of the potential successor are all important factors in determining
their suitability for the role.

Furthermore, it is critical that the chosen successor be aligned with the company's
culture, values, and strategic direction, as well as capable of maintaining and
improving the company's reputation and stakeholder relationships.

Return also one of the critical component of the succession plan because it influences the successor's motivation, commitment, and loyalty to the company. The deal's
terms should be competitive and in line with industry norms, while also reflecting
the successor's contributions and performance.

To summarize, the succession plan must be well-planned and implemented, taking
into account the family business's unique characteristics and needs. To ensure a
smooth transition of power and business continuity, it should include open communication and collaboration among current and potential business leaders, family
members, and other stakeholders.

References

1. Bernama, BNM : Ekonomi Malaysia Menguncup 17 (2020). https://www.benarnews.org/malay/berita/my-bnm-200814-08142020175822.html
2. H. Metro, Peniaga Kecil Sesuaikan Diri Norma Baharu (2020). https://www.hmetro.com.my/mutakhir/2020/04/571425/peniaga-kecil-sesuaikan-diri-norma-baharu-metrotv
3. Institute for Family Business, Succession planning in family companies (2020). https://www.ifb.org.uk/resources/succession-planning-in-family-business/
4. T.T. Pham, R. Bell, D. Newton, The father's role in supporting the son's business knowledge development process in Vietnamese family businesses. J. Entrep. Emerg. Econ., 11(2), 258–276 (2019). https://doi.org/10.1108/JEEE-01-2018-0006
5. R. Bell, T.T. Pham, Modelling the knowledge transfer process between founder and successor in Vietnamese family businesses succession. J. Fam. Bus. Manag. (2020). https://doi.org/10.1108/JFBM-03-2020-0024
6. E.-O.-I. Lucky, M.S. Minai, A.O. Isaiah (2011) A conceptual framework of family business succession: bane of family business continuity. Int. J. Bus. Soc. Sci.
7. P. Englisch, F. Ambrosini (2020) Family Businesses and COVID-19: Managing Ownership: The four Key Areas That Family Business Owner Should Consider. PwC
8. S. Rashid, V. Ratten, A dynamic capabilities approach for the survival of Pakistani family-owned business in the digital world. J. Fam. Bus. Manag. (2020). https://doi.org/10.1108/JFBM-12-2019-0082
9. T. Nyoni, Factors affecting succession planning in small and medium enterprises (SMEs) in Zimbabwe: a case study of Harare. MPRA (2019)
10. V. Ramadani, R.D. Hisrich, L.P. Dana, R. Palalic, L. Panthi, Beekeeping as a family artisan entrepreneurship business. Int. J. Entrep. Behav. Res. (2019) https://doi.org/10.1108/IJEBR-07-2017-0245
11. P. Sandu, N. Nye, Succession challenges in family businesses from the first to the second generation. J. Small Bus. Entrep. Dev. (2020). https://doi.org/10.15640/jsbed.v8n1a5
12. K. Kandade, G. Samara, M.J. Parada, A. Dawson, From family successors to successful business leaders: a qualitative study of how high-quality relationships develop in family businesses. J. Fam. Bus. Strategy. (2020). https://doi.org/10.1016/j.jfbs.2019.100334
13. A. Mosbah, J. Alharbi, Family business research in Malaysia: thematic and methodological assessment. Int. J. Manag. (2020). https://doi.org/10.34218/IJM.11.11.2020.057
14. A. Mosbah, S.R. Serief, K.A. Wahab. Performance of family business in Malaysia. Int. J. Soc. Sci. Perspect. (2017) https://doi.org/10.33094/7.2017.11.20.26
15. W. Dusor, *Succession Plan in Family Business: A Case of Ketu South Municipality*, University of Cape Coast (2020)
16. A.S. Bathija, R.G. Priyadarshini, A study on factors affecting succession planning in small and medium scale Indian family business. IOP Conf. Ser. Mater. Sci. Eng. (2018) https://doi.org/10.1088/1757-899X/390/1/012091
17. A. Mohamad, N. Naiimi, S. Abdullah, Memimpin pengambilalihan perniagaan keluarga: kaedah, masalah dan cara mengatasinya. Int. J. Manag. Stud. (2018)
18. V. Koráb, Key factors influencing family businesses: a qualitative study of selected world wineries in. SHS Web Conf. (2021). https://doi.org/10.1051/shsconf/202111502003
19. J.F. LeCounte, Founder-CEOs: succession planning for the success, growth, and legacy of family firms. J. Small Bus. Manag. (2020). https://doi.org/10.1080/00472778.2020.1725814
20. K. Randerson, C. Seaman, J.J. Daspit, C. Barredy. Institutional influences on entrepreneurial behaviours in the family entrepreneurship context: towards an integrative framework. Int. J. Entrep. Behav. Res. (2020). https://doi.org/10.1108/IJEBR-01-2020-824
21. T.S. Chin, A. Jusoh, Sejarah Penglibatan Orang Cina Dalam Perniagaan Keluarga (Family Business) Berasaskan Industri Penghasilan Makanan Di Butterworth, 1957–2000 (2021)
22. S. Tehseen, A.R. Anderson, Cultures and entrepreneurial competencies; ethnic propensities and performance in Malaysia. J. Entrep. Emerg. Econ. (2020). https://doi.org/10.1108/JEEE-10-2019-0156

23. B. Ten Holte, *Making Family Business Succession into Family Business Success*. Radbound University (2019)
24. S. Teixeira, P. Mota Veiga, R. Figueiredo, C. Fernandes, J.J. Ferreira, M. Raposo, A systematic literature review on family business: insights from an Asian context. J. Fam. Bus. Manag. (2020). https://doi.org/10.1108/JFBM-12-2019-0078
25. Kementerian Pembangunan Usahawan. Dasar Keusahawanan Nasional 2030 (2019)
26. D. Pittino, F. Chirico, M. Baù, M. Villasana, E.E. Naranjo-Priego, E. Barron, Starting a family business as a career option: the role of the family household in Mexico. J. Fam. Bus. Strateg. (2020). https://doi.org/10.1016/j.jfbs.2020.100338
27. G. Leiß, A. Zehrer, Intergenerational communication in family firm succession. J. Fam. Bus. Manag. (2018). https://doi.org/10.1108/JFBM-09-2017-0025
28. M. Dayan, P.Y. Ng, N.O. Ndubisi, Mindfulness, socioemotional wealth, and environmental strategy of family businesses. Bus. Strateg. Environ. (2019). https://doi.org/10.1002/bse.2222
29. M. Wu, M. Coleman, A.R. Abdul Rahaman, B.K. Edziah, Successor selection in family business using theory of planned behaviour and cognitive dimension of social capital theory: evidence from Ghana. J. Small Bus. Enterp. Dev. (2020). https://doi.org/10.1108/JSBED-05-2019-0152
30. J. Buckman, P. Jones, S. Buame, Passing on the baton: a succession planning framework for family-owned businesses in Ghana. J. Entrep. Emerg. Econ. (2019). https://doi.org/10.1108/JEEE-11-2018-0124
31. E.L. Giménez, J.A. Novo, A theory of succession in family firms. J. Fam. Econ. (2020). https://doi.org/10.1007/s10834-019-09646-y
32. S. Savolainen (2020) Psychodynamics from the Perspective of Non-family Employees during a Small Family Business Succession. Univers. J. Manag. https://doi.org/10.13189/ujm.2020.080401
33. L.K. Fisher, *Family Owned Business: A Case Study of First Through Third Generation Succession Planning*. Capella University (2017)
34. A. Cherdchai, *Entrepreneurial Learning in Multigenerational Family Business Succession*. University of Westminster (2020)
35. M. Gagné, C. Marwick, S. Brun de Pontet, C. Wrosch, Family business succession: what's motivation got to do with it? Fam. Bus. Rev. (2019). https://doi.org/10.1177/0894486519894759
36. S. Zhu, The influence of incumbent- and successor-related factors on succession process in family firms: Exploring the role of values in family business succession. Singapore Management University (2020)
37. J.K.K. Tang, W.S. Hussin, Next-generation leadership development: a management succession perspective. J. Fam. Bus. Manag. (2020). https://doi.org/10.1108/JFBM-04-2019-0024
38. F. Lu, H.K. Kwan, Z. Zhu, The effects of family firm CEO traditionality on successor choice: the moderating role of socioemotional wealth. Fam. Bus. Rev. (2021). https://doi.org/10.1177/0894486520967832
39. C. Matias, M. Franco, The role of the family council and protocol in planning the succession process in family firms. J. Fam. Bus. Manag. (2020). https://doi.org/10.1108/JFBM-01-2020-0004
40. M.R. Razzak, R. Abu Bakar, N. Mustamil, Socioemotional wealth and family commitment: moderating role of controlling generation in family firms. J. Fam. Bus. Manag. (2019). https://doi.org/10.1108/JFBM-09-2018-0050
41. D. Gimenez-Jimenez, L.F. Edelman, T. Minola, A. Calabrò, L. Cassia, An intergeneration solidarity perspective on succession intentions in family firms. Entrep. Theory Pract (2021). https://doi.org/10.1177/1042258720956384
42. K. Eddleston, I.C. Botero, Team exercise: how parenting styles affect the next generation. Entrep. Innov. Exch. (2021). https://doi.org/10.32617/601-6048f904eab23
43. I. Umans, N. Lybaert, T. Steijvers, W. Voordeckers, Succession planning in family firms: family governance practices, board of directors, and emotions. Small Bus. Econ. (2020)
44. S. Schell, J.K. de Groote, P. Moog, A. Hack, Successor selection in family business—a signaling game. J. Fam. Bus. Strateg. (2020). https://doi.org/10.1016/j.jfbs.2019.04.005

45. I.H. Bokhari, A.B. Muhammad, N. Zakaria, Succession planning, strategic flexibility as predictors of business sustainability in family-owned SMEs: moderating role of organization improvisation. Pakistan J. Commer. Soc. Sci. (2020)
46. N. Udomkit, P. Kittidusadee, C. Schreier, Disharmony within harmony: contrasting views between incumbents and successors on the selection criterion adopted for family business successions. J. Fam. Bus. Manag. (2021)
47. C. Bulut, S. Kahraman, E. Ozeren, S. Nasir, The nexus of aging in family businesses: decision-making models on preferring a suitable successor. J. Organ. Chang. Manag. (2019). https://doi.org/10.1108/JOCM-05-2019-0140
48. F. M. Adly, G. Anggadwita, Analisis Succession planning Pada family business Berbudaya Tionghoa di Kota Bandung. e-Proceed. Manag. (2018)
49. A. Calabrò, A. Minichilli, M.D. Amore, M. Brogi, The courage to choose! Primogeniture and leadership succession in family firms. Strateg. Manag. J. (2018). https://doi.org/10.1002/smj.2760
50. H. Fürst, *Growing Up in a Business Family—An Analytic Autoethnography of 'Subtle Coerced Succession*. The University of Gloucestershire (2017)
51. G. Osnes, L. Hök, O. Yanli Hou, M. Haug, V. Grady, J.D. Grady, Strategic plurality in inter-generational hand-over: incubation and succession strategies in family ownership. J. Fam. Bus. Manag. (2019). https://doi.org/10.1108/JFBM-06-2018-0018
52. A. Kubíček, O. Machek, Gender-related factors in family business succession: a systematic literature review. Rev. Manag. Sci. (2019). https://doi.org/10.1007/s11846-018-0278-z
53. L.K.B. Martini, I.G.A.M.D. Dewi, The effect of successor characteristics on succession planning of family companies. J. Appl. Manag. (JAM) (2020)
54. J. Zybura, N. Zybura, J.P. Ahrens, M. Woywode, Innovation in the post-succession phase of family firms: family CEO successors and leadership constellations as resources. J. Fam. Bus. Strateg. (2020). https://doi.org/10.1016/j.jfbs.2020.100336
55. M.A. Abdullah, Z. Abdul Hamid, J. Hashim, Family-owned businesses: towards a model of succession planning in Malaysia. Int. Rev. Bus. Res. Pap. 7(1), 251–264 (2011)
56. C. Magasi, Investigation of successor selection determinants and their effect on family business survival. African J. Appl. Res. (2021). https://doi.org/10.26437/ajar.10.2021.05
57. P. Liamputtong, *Qualitative Research Methods* (Oxford University Press, 2019)
58. J.W. Creswell, *Research Design: Qualitative, Quantitative, and Mixed Methods Approaches*, 4th edn. (SAGE Publications, Thousand Oaks, California, 2014)

Using Data to Enhance Higher Education in the Age of IR 4.0: A Rapid Scoping Review

Jawahir Che Mustapha, Munaisyah Abdullah, Husna Osman, and Husna Sarirah Husin

Abstract Higher education institutions are facing challenges to prepare students for the rapidly changing world of work in the era of Industry 4.0. At the same time, these institutions have access to unprecedented amounts of data that can help them better understand student needs and outcomes, as well as the effectiveness of teaching and learning strategies. This paper used a rapid scoping review to explore the current landscape of data utilization in higher education, specifically in terms of how data is being used to inform decision-making and improve outcomes. The findings suggest that data utilization in higher education requires more attention from researchers and practitioners to fully realize its potential. This paper concludes by discussing the implications of the findings and suggesting areas for future research.

Keywords Data utilization · Rapid scoping review · Higher education

1 Introduction

In the age of the fourth industrial revolution (IR 4.0), higher education institutions (HEIs) are increasingly turning to data-driven decision-making (DDDM) to improve their operations and outcomes. Education 4.0 is a response to the changes brought

J. C. Mustapha (✉) · M. Abdullah · H. Osman · H. S. Husin
Informatics and Analytics Section, Universiti Kuala Lumpur Malaysian Institute of Information Technology, Kuala Lumpur, Malaysia
e-mail: jawahir@unikl.edu.my

M. Abdullah
e-mail: munaisyah@unikl.edu.my

H. Osman
e-mail: husna@unikl.edu.my

H. S. Husin
e-mail: sarirah@unikl.edu.my

© The Author(s), under exclusive license to Springer Nature Switzerland AG 2024
A. Ismail et al. (eds.), *Tech Horizons*,
SpringerBriefs in Applied Sciences and Technology,
https://doi.org/10.1007/978-3-031-63326-3_10

about by IR 4.0 and represents a shift toward a more personalized, technology-enabled, and lifelong approach to learning that is better suited to the needs of the modern world. In this context, DDDM has become increasingly important for HEIs to remain competitive and relevant in the global knowledge economy [1].

Data utilization, defined as the continuous use of data to inform decision-making [2], is a key aspect of DDDM in HEIs. DDDM has become a crucial aspect of higher education management, as it provides valuable insights and information to make informed decisions. Data can be generated from various sources, such as student information systems, learning management systems, social media, and Internet of things (IoT) devices, and can be analyzed using advanced techniques such as machine learning, artificial intelligence, and data visualization.

Data utilization in HEIs is becoming increasingly crucial in the age of IR 4.0, as the sector seeks to harness the power of data to inform decision-making and improve outcomes. Despite a growing number of studies on the subject, there is still much to be understood about the current landscape of data utilization and its potential impact on the sector. Therefore, the purpose of this paper is to explore the current landscape of data utilization in HEIs in the age of IR 4.0, focusing on how data is being used to inform decision-making and improve outcomes. By offering insights into the ways in which data is being used to inform decision-making and improve outcomes, this research can play a significant role in advancing the understanding of data utilization in HEIs and can contribute to the development of best practices in this field and also to motivate similar new zone research in future. The intended audience for this accessible overview includes individuals who may not have a strong background in data utilization in HEIs but are interested in learning more about this important topic. This paper aims to encourage these individuals to further explore this area of research and practice, and to contribute to the ongoing efforts to enhance higher education in the age of IR 4.0.

The rest of the paper is organized as follows: Sect. 2 explains the methodology, which adopts a rapid scoping review approach. Section 3 reports the findings of research questions. Section 4 presents the discussion and conclusion and research implications. Finally, conclude with a summary of the main findings and recommendations for future research and practice.

2 Methodology

The Arksey and O'Malley's framework [3] was modified to prioritize early data extraction and synthesis, allowing for a faster timeline in scoping reviews and quicker identification of trends in the literature which involves the following stages:

1. Research Question Identification: Specifically, this review seeks to answer the following research questions: (i) What is data utilization in the realm of data-driven decision-making in higher education? (ii) What types of data are being

collected and analyzed in higher education? and (iii) How is the data being used to inform decision-making and improve outcomes in higher education?

2. Search Strategy: A search of electronic databases including Google Scholar, IEEE, ERIC, and Scopus was conducted to identify relevant studies. The search terms used were "data utilization" OR "data-driven decision-making" AND "higher education," OR "university" OR "college" AND "Industry 4.0".

3. Study Selection: To be included in this review, studies had to meet the following criteria: (i) Written in English language, (ii) Focus on the use of data in higher education for decision-making, and (iii) Written between 2018 and 2023.

4. Charting the Data: Data was extracted from the studies using a standardized form, including study characteristics, data types, and outcomes related to data use in higher education. Thematic analysis was applied to the extracted data to identify themes related to the research questions: data utilization, types of data, and impact of data-driven decision-making on outcomes in higher education.

5. Collating, Summarizing, and Reporting the Results: The extracted data were synthesized and summarized in narrative form, organized by research question.

3   Results

After removing duplicates and applying the inclusion and exclusion criteria, 25 articles were included in this scoping review, and this section of the study discusses the research questions by illustrating relevant studies included in the review.

3.1   *What is Data Utilization in the Realm of Data-Driven Decision-Making in Higher Education?*

The reviewed literature did not provide a clear definition of data utilization in the context of data-driven decision-making in HEIs. However, it can be understood as the practice of collecting and leveraging various types of student data from multiple sources to inform decision-making, communication tactics, and support efforts in HEIs [4]. This approach allows universities to adopt a data-driven approach to education and extract valuable insights into the teaching and learning process [5]. Analytics are crucial in this context as they facilitate performance optimization and enable data driven decision-making [5]. It is important to note that data management should be driven by data utilization and that data should be used only as appropriate to support university missions and activities [6].

3.2   What Types of Data Are Being Collected and Analyzed in Higher Education?

The studies reported on a wide range of data types that are collected and analyzed in HEIs, including academic performance data such as growth, courses taken, grades, number of credits earned, and degree progress [7–13], as well as demographic data including age, gender, ethnicity, race, socioeconomic status, postcode, and special education needs [7, 9–16].

Other data categories included engagement [8], interests [7, 9], attendance [12, 13, 15], behavior such as actions and interactions [7, 8, 13], historical academic performance data such as school grades and test scores [16–18], assessment data including quiz, test, e-portfolio, and online course data [15, 19–21], feedback and evaluation data [7, 22], extracurricular activity data [7], program participation data [7], network profile data [14], tracer study data such as job titles, salary levels, and industry sectors [23], department/faculty data [10], and financial data such as budgets, expenses, and revenues [13].

3.3   How is the Data Being Used to Inform Decision-Making and Improve Outcomes in Higher Education?

The studies reviewed reported a range of ways in which data is being used to inform decision-making and improve outcomes in higher education. By analyzing academic performance, online activities attendance, demographic, behavioral, and social-emotional data, HEIs can gain insights into student performance and behavior, develop predictive models, early warning system and identify students who may be at risk of falling behind. This enables HEIs to develop targeted interventions and personalized learning plans for academic advising, help all students achieve academic success and improve graduation rates [9–13, 15–17, 24, 25]. Furthermore, by analyzing institutional data, social media news, and tracer study data, institutions, educators, and career counselors can gain a better understanding of curriculum relevance, students' performance, and career interests. This, in turn, can help guide them toward career paths that are most suitable for their skills, interests, and goals [7, 23].

Through the analysis of student data, educators can enhance teaching and learning outcomes by identifying challenging areas for students, adjusting instructional approaches, and providing specific assistance to promote student learning. This process enables educators to improve the educational experience for students by tailoring their teaching strategies to meet individual needs [8, 26]. Educators can also leverage student records from university databases and network profiles from telecommunication towers to develop applications such as a decision-making dashboard for top management and a system for lecturers to plan teaching approaches based on students' internet accessibility. This enables a data-driven approach to teaching and learning that improves the online educational experience for students [14]. In addition, by analyzing e-portfolios as well as students' feed-back, educators can gain insights into student learning and engagement, identify patterns and trends, and evaluate the effectiveness of teaching methods and optimize the curriculum. Ed-tech companies also use analytics to help students track their progress and identify areas where they need additional support, while providing insights into user experience to support student learning [19–22].

Furthermore, financial data analysis can help HEIs identify patterns and trends in their financial performance, detect potential financial risks, and make informed decisions to improve their financial management practices. Through the use of early warning models, HEIs can be better prepared to mitigate financial risks and improve their overall financial stability [27]. By analyzing data on student demographics, interests, and academic performance, HEIs can gain valuable insights into the characteristics of their most successful recruits and identify potential recruits who are a good fit for the HEI. This data can inform the development of targeted marketing strategies, personalized communications and outreach efforts, and other recruitment and admissions initiatives that are more likely to resonate with prospective students and ultimately lead to increased success in recruitment and enrollment [18, 28].

4   Discussion

This study has identified the current landscape of data utilization in HEIs including the ways in which data is being used to inform decision-making and improve outcomes. The findings of this review suggest that data-driven decision-making is becoming increasingly prevalent in HEIs, and that it has the potential to improve institutional performance and outcomes in a variety of ways.

4.1   Data Utilization in Higher Education

It is evident that there is a lack of a clear definition of data utilization in the context of data-driven decision-making in HEIs. However, it can be understood as the continuous process of collecting, analyzing, and using data to make informed decisions, improve operational efficiency and enhance outcomes related to teaching, learning, research, and administration in HEIs. It involves using various types of data to inform decision-making and to measure progress toward institutional goals. The purpose of data utilization in higher education is to leverage data to improve educational activities. It applies the insights gained from data analysis to drive action and improve outcomes.

Data utilization and data analysis are interrelated processes that depend on each other for success. Data utilization involves using data to inform decision-making, monitor progress, and assess outcomes to improve the quality and effectiveness of education. This process involves collecting and analyzing data from various sources such as student records, course evaluations, and institutional assessments to identify patterns, trends, and insights. These insights are then used to develop strategies, policies, and practices to improve student learning outcomes, institutional effectiveness, and efficiency.

Data analysis is a critical component of data utilization, as it involves examining and interpreting data to identify patterns, trends, and insights. Effective data analysis provides the evidence necessary for informed decision-making and action-taking. It helps HEIs to identify areas for improvement, evaluate the effectiveness of existing programs or interventions, and track progress toward institutional goals. Data analysis in HEIs involves applying various statistical and analytical techniques to understand the relationships between different variables and to identify patterns and trends.

Once insights are gained through data analysis, data utilization involves applying those insights to make informed decisions and take action to drive institutional goals. For instance, if data analysis reveals that a certain program is not achieving the desired outcomes, data utilization might involve making changes to the program to improve its effectiveness. Data utilization and data analysis are cyclical processes that continually inform and build upon each other. As HEIs implement changes based on data analysis, they collect new data which can then be used for further analysis and improvement. This ongoing cycle of data utilization and analysis helps HEIs to continually improve and adapt to changing circumstances.

4.2 *Implications for Practice*

The studies reviewed in this scoping review reported a range of outcomes associated with the use of data-driven decision-making in HEIs. The use of data in higher education can provide valuable insights that can inform decision-making and ultimately lead to improved outcomes for students. HEIs can use data to identify at-risk students and provide targeted interventions to improve academic performance and graduation rates. Educators can use data to tailor their teaching strategies to meet individual student needs and improve the educational experience. Admissions and recruitment efforts can be informed by data to better identify and attract suitable students to the HEI. Financial management can also benefit from the use of data to identify potential risks and make informed decisions.

However, it is important to recognize the ethical and privacy concerns that come with the collection and use of student data. HEIs must ensure that data is collected and used in a responsible manner, and that students' privacy rights are protected. Additionally, HEIs should also consider the potential limitations of data-driven decision-making and recognize that it should not replace the professional judgment and expertise of educators and administrators.

4.3 Future Research

As the use of data becomes more prevalent in higher education, there are several areas for future research to explore. First, ethical considerations and privacy concerns associated with collecting and using student data should be explored. Additionally, the impact of data-driven decision-making on teaching and learning outcomes should be evaluated. Supporting underrepresented students, leveraging data for lifelong learning and career development, and using data for institutional effectiveness and strategic planning are also potential areas for future research. By addressing these areas, HEIs can ensure that the use of data is ethical, effective, and beneficial for all students.

5 Conclusion

This study provides an up-to-date overview of the current state of data utilization in higher education, which can serve as a reference for educators and administrators interested in leveraging data to inform decision-making. The scoping review methodology used in this study can be replicated in future research to explore other areas related to data utilization in higher education. This study identifies gaps in the existing literature and suggests areas for future research, which can help to advance the field and improve the effectiveness of data utilization in higher education. As higher education continues to evolve in the age of IR 4.0, it will be important for HEIs to embrace data-driven decision-making as a tool for improving institutional performance and outcomes and ensuring the success of their students in an ever-changing world.

Acknowledgements Acknowledgment is made for the grant (STR21040) received from UniKL in support of this work. Their generous support is greatly appreciated.

References

1. A. Fernández, B. Gómez, K. Biniaku, E.K. Meçe, Digital transformation initiatives in higher education institutions: a multivocal literature review. Educ. Inf. Technol. 1–32 (2023)
2. Medical.Data.Vision.: What is Data Utilization? Introducing the Benefits and How it can be Utilized in Different Industries|Medical Data Vision Co., Ltd. tinyurl.com/4d8w7fv5 (2021). Accessed 2 March 2023
3. H. Arksey, L. O'Malley, Scoping studies: towards a methodological framework. Int. J. Soc. Res. Methodol. **8**(1), 19–32 (2005)
4. J. Dunlap, I. Palmer, A. Wesley, *Keeping Student Trust: Student Perceptions of Data Use within Higher Education*, New America (2021)
5. S. Mokhtar, J.A. Alshboul, G.O. Shahin, Towards data-driven education with learning analytics for educator 4.0. J. Phys. Conf. Series (2019)

6. University of Delaware.: Managing data utilization. https://www1.udel.edu/security/data/utiliz ation.html (n.d.). Accessed 4 Feb 2023
7. T.C. Yang, C.Y. Chang, Using Institutional data and messages on social media to predict the career decisions of university students—a data-driven approach. Educ. Inf. Technol. **28**(1), 1117–1139 (2023)
8. D. Ifenthaler, C. Schumacher, J. Kuzilek, Investigating students' use of self-assessments in higher education using learning analytics. J. Comput. Assist. Learn. **39**(1), 255–268 (2023)
9. R. Wang, J.E. Orr Jr., Use of data analytics in supporting the advising of undecided students. J. Coll. Stud. Retention Res. Theory Pract. **23**(4), 824–849 (2022)
10. M. Yağcı, Educational data mining: prediction of students' academic performance using machine learning algorithms. Smart Learn. Environ. **9**(1), 11 (2022)
11. H.T.H. Duong, L.T.M. Tran, H.Q. To, K. Van Nguyen, Academic performance warning system based on data driven for higher education. Neural Comput. Appl. 1–19 (2022)
12. A. Hershkovitz, A. Ambrose, Insights of instructors and advisors into an early prediction model for non-thriving students. J. Learn. Anal. **9**(2), 202–217 (2022)
13. O. Moscoso-Zea, P. Saa, S. Luján-Mora, Evaluation of algorithms to predict graduation rate in higher education institutions by applying educational data mining. Australas. J. Eng. Educ. **24**(1), 4–13 (2019)
14. N.O. Haryani Haron, S. Aliman, A. Ismail, M.Y. Darus, A.M. Ariffin, Using data-driven for improved educational experience during covid19. J. Positive Sch. Psychol. **6**(3), 8599–8610 (2022)
15. S. Gaftandzhieva, A. Talukder, N. Gohain, S. Hussain, P. Theodorou, Y.K. Salal, R. Doneva, Exploring online activities to predict the final grade of student. Mathematics **10**(20), 3758 (2022)
16. P.D. Gil, S. da Cruz Martins, S. Moro, J.M. Costa, A data-driven approach to predict first-year students' academic success in higher education institutions. Educ. Inf. Technol. **26**(2), 2165–2190 (2021)
17. R. Bütüner, M.H. Calp, Estimation of the academic performance of students in distance education using data mining methods. Int. J. Assess. Tools Educ. **9**(2), 410–429 (2022)
18. H.A. Mengash, Using data mining techniques to predict student performance to support decision making in university admission systems. IEEE Access, 8, 55462–55470 (2020)
19. A. Allman, A. Kocnevaite, F. Nightingale, The effectiveness of online portfolios for assessment in higher education, in *The IAFOR International Conference on Education*, Hawaii (2021)
20. M.F. Van der Schaaf, Electronic portfolios enhanced with learning analytics at the workplace. *Handbook of Vocational Education and Training*, 1–20 (2019)
21. C. Li, N. Herbert, S. Yeom, J. Montgomery, Retention factors in STEM education identified using learning analytics: a systematic review. Educ. Sci. **12**(11), 781 (2022)
22. Y. Du, Application of the data-driven educational decision-making system to curriculum optimization of higher education. Wirel. Commun. Mob. Comput. 1–8 (2022)
23. M. Arifin, Using education data mining (EDM) and tracer study (TS) data as materials for evaluating higher education curriculum and policies. KnE Soc. Sci. 26–35 (2022)
24. K. Nahar, B.I. Shova, T. Ria, H.B. Rashid, A.S. Islam, Mining educational data to predict students' performance: a comparative study of data mining techniques. Educ. Inf. Technol. **26**(5), 6051–6067 (2021)
25. O. Iatrellis, I.K Savvas, P. Fitsilis, V.C. Gerogiannis, A two-phase machine learning approach for predicting student outcomes. Educ. Inf. Technol. **26**, 69–88 (2021)
26. M. Usher, A. Hershkovitz, A. Forkosh-Baruch, From data to actions: instructors' decision making based on learners' data in online emergency remote teaching. Br. J. Edu. Technol. **52**(4), 1338–1356 (2021)
27. Z. Zang, Analysis of financial management and decision-making in institution of higher learning based on deep learning algorithm. *Mobile Information Systems* (2022)
28. M. Kurzweil, M. Stevens, Setting the table: responsible use of student data in higher education. EDUCAUSE. tinyurl.com/32fpa6d6 (2018). Accessed 4 Feb 2023

Prediction Model of Cardiovascular Diseases Using ANFIS Sugeno

Sri Sumarlinda, Azizah Binti Rahmat, and Zalizah Awang Long

Abstract Cardiovascular diseases are among the killer diseases in the world. The diseases caused a lot of death, and disabilities, and contributed to high costs of treatment. Early treatment of cardiovascular diseases by knowing the risk factors for disease susceptibility will facilitate treatment and healing. This study aims to develop a cardiovascular disease prediction model using the Sugeno's adaptive neuro-fuzzy inference system (ANFIS). The grid partition and sub-clustering were used in the developed ANFIS. The data set comprises clinical data from UCI Global Data. The result analysis of the prediction model generated Fuzzy Inference System (FIS) uses grid partition with optimization backpropagation, grid partition with optimization hybrid, sub-clustering with optimization backpropagation, sub-clustering with optimization hybrid values of root mean square error are 0.7059, 0.2579, 0.7071, and 0.2576, respectively.

Keywords Prediction · Cardiovascular diseases · ANFIS · Grid partition · Sub-clustering

S. Sumarlinda (✉) · A. B. Rahmat · Z. A. Long
Universiti Kuala Lumpur Malaysian Institute Information Technology, Kuala Lumpur, Malaysia
e-mail: sumarlinda.sri@s.unikl.edu.my

A. B. Rahmat
e-mail: azizah@unikl.edu.my

Z. A. Long
e-mail: zalizah@unikl.edu.my

S. Sumarlinda
Faculty of Computer Science, Duta Bangsa University, Surakarta, Indonesia

© The Author(s), under exclusive license to Springer Nature Switzerland AG 2024
A. Ismail et al. (eds.), *Tech Horizons*,
SpringerBriefs in Applied Sciences and Technology,
https://doi.org/10.1007/978-3-031-63326-3_11

1 Introduction

Cardiovascular diseases are arising both globally and nationally. Many individuals and countries spend significant funds to prevent it. The growth of disease and its influence on personal and community life is concerning [1–3]. Identification and diagnosis of a disease facilitate the healing and recovery of patients. Early identification of cardiovascular disease by knowing the risk factors for cardiovascular disease susceptibility reduces the cost and duration of healing and makes it easier to treat. Cardiovascular risk factors are diverse and numerous and develop following the development of the human lifestyle. Gender is a risk factor for cardiovascular disease and is related to other risk factors such as obesity and physical activity. In Saudi Arabia, women tend to be more obese and lack physical activity [3]. The vulnerability identification for cardiovascular disease prevention is also carried out by studying social demographics (gender, race, obesity, etc.), which are then used for prevention by improving lifestyle, physical activities, and health responsibilities [2]. Body mass index measurements, smoking state, lifestyles, and genetic factors are identified as risk factors for cardiovascular diseases [3, 4].

Prediction and classification are data mining functions whose applications are extensively used in various fields of life. The accuracy and power of the prediction model are essential in its application. Accuracy and robustness are the main concerns for improvement. Prediction and classification models are facilitated by machine learning algorithms such as support vector machines, linear regression, artificial neural networks, and ANFIS [5]. The problem in improving the performance of prediction and classification models is the difficulty of convergence so that they are stuck at local optima and the effects of gradient descent. Various efforts to provide solutions to convergence and gradient descent problems have been carried out, for example, by adding learning parameters such as momentum, jumping steps such as residual and resilient networks, adding layers to the process, modifying the learning process, and so on [5–8].

The advantage of FIS can approximate human reasoning abilities and thereby develop inference mechanisms based on IF–THEN rules from knowledge-based systems (data input–output or human expertise) and robust mathematical models with intrinsic capabilities to deal with the inevitable uncertainties associated with processes. The weakness of FIS is it cannot learn and adapt to the environment to produce output within the required error rate. The mechanism of ANFIS is close to ANN, which can adapt and learn from data. The combination of FIS and ANN creates the potential for better accuracy as a comprehensive system, and ANFIS also eliminates or mitigates certain drawbacks associated with both methods. For example, it has been shown that ANFIS reduces the need for an expert. ANFIS is a powerful tool for building complex and nonlinear relationships between input–output data for function estimation and pattern classification [7–9].

The capabilities of ANFIS combine the advantages of system/logic fuzzy models and artificial neural networks. The ANFIS can extract significant features from data using backpropagation learning algorithms and least squares algorithms. ANFIS is highly flexible and adaptable to model various data types, including numerical, categorical, and ordinal scale data. ANFIS can produce predictive models that are accurate and intuitively interpretable, making it easier to make decisions based on the prediction results [1, 2]. ANFIS can be used in various fields such as environmental science, finance, health, and engineering, so this method is comprehensive in its application [3–6]. ANFIS is also easy to implement in a software like MATLAB, making it easier for users to build and implement predictive models quickly and efficiently [7]. The ANFIS is a network with a radial position to be equivalent to fuzzy rules based on the first-order Sugeno model [10, 11].

The purpose of this study is to analyze the prediction model of cardiovascular disease. The prediction model generates a FIS grid partition and sub-clustering optimization method using backpropagation and hybrid (combination of backpropagation and least square algorithm). The model development consists of four steps: grid partition backpropagation, grid partition hybrid, sub-clustering backpropagation, and sub-clustering hybrid.

2 Methodology

The study embarked on two approaches which are descriptive and explorative research. The research method consists of three primary stages, namely: (1) Data collecting and data processing; (2) Model development; and (3) Model evaluation. The details of each stage are depicted in Fig. 1.

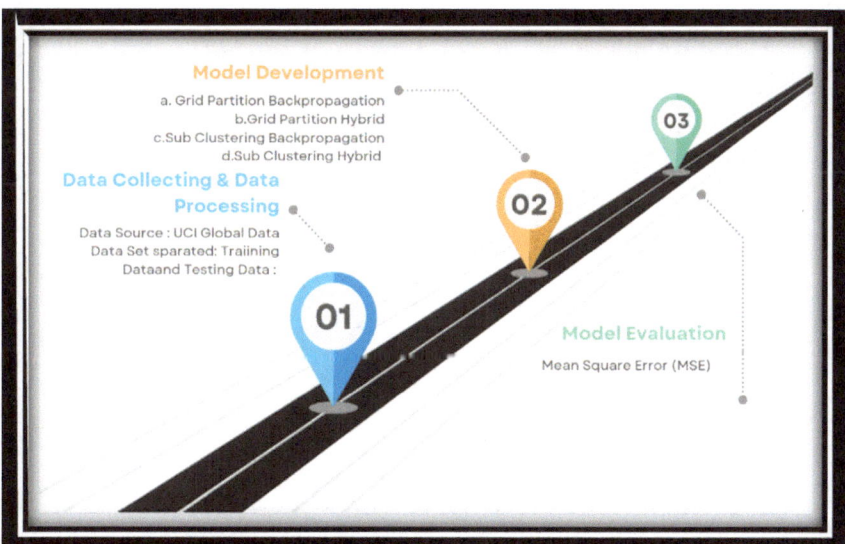

Fig. 1 Flow diagram of the research method

2.1 Data Collecting and Data Processing

The data set is from UCI global data. The number of data sets is 461 data (sample of a patient). The variables in the data set are id, age, sex, cp, trest bps, chol, fbs, restech, thalach, exang, oldpeak, slope, and cardio. Based on the variables, id is the attribute identifier while cardio is the label and other variables are the predictive attributes.

The data set consists of 70% training purpose data (about 323 samples) and 30% testing data (about 138). The 461 samples consist of the label 1 value (cardiovascular diseases) 261 samples (57%) and 0 value (not cardiovascular disease) 200 samples (43%). For training purposes, data were separated similarly to the data set composition of 57% for the value 1 label (184 samples) and for the value 0 label about 43% (133 samples). The number of testing data is 138 samples.

2.2 Model Development

This study used ANFIS Sugeno to analyze the prediction model for cardiovascular diseases. The Sugeno's fuzzy model (fuzzy model TSK) was proposed by Takagi, Sugeno, and Kang attempted to construct a systematic approach to generate fuzzy rules of a given input–output data set [12]. The Sugeno ANFIS model is shown in Fig. 2. The optimization method used backpropagation and a hybrid algorithm.

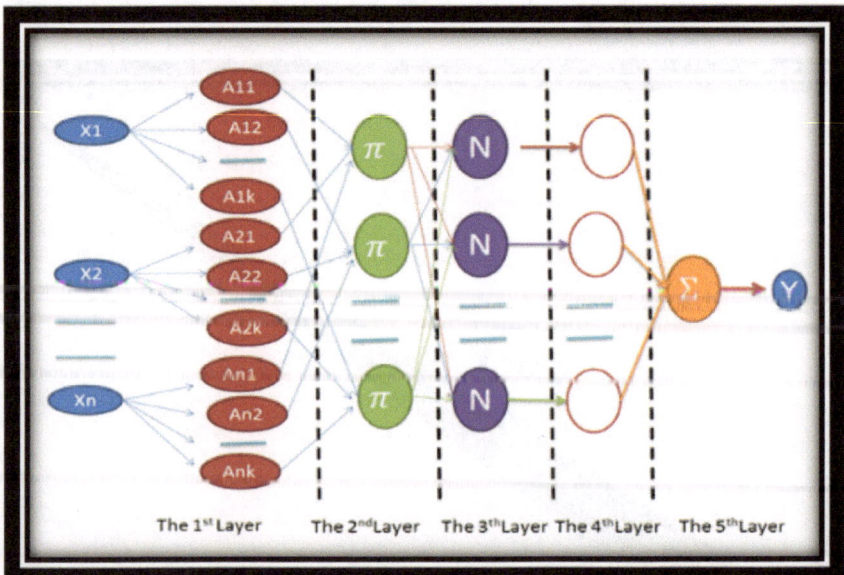

Fig. 2 The Sugeno ANFIS model

(a) Grid Partition

The FIS grid partition is commonly used in applications such as control systems, where the input variables have continuous ranges of values, and a finite number of rules need to be defined to control the system. The FIS grid partition technique is useful for complex systems where a large number of rules are required to control the system, and it provides a way to simplify the rule base by reducing the number of rules required to describe the system [12–16].

(b) Sub-clustering

For the ANFIS sub-clustering, the input data is first partitioned into sub-clusters using a fuzzy c-means algorithm, which assigns each data point a membership value indicating its degree of belonging to each sub-cluster. Then, a neural network model is trained on the sub-clustered data using backpropagation to learn the mapping between the input data and the output data. However, it can also be more computationally expensive and difficult to interpret than other clustering methods [10–12, 17, 18].

(c) Backpropagation

For backpropagation, the neural network uses an input data point to make a prediction and then compares the prediction to the actual output. The difference between the prediction and the actual output is known as the error or loss. The backpropagation algorithm calculates the gradient of the loss with respect to the weights of the neural network and uses this gradient to update the weights in order to reduce the error. This process is repeated for many iterations until the neural network's predictions become accurate enough. The algorithm gets its name from the way it works backwards through the network, starting from the output layer and propagating the error backwards through the layers of the network, updating the weights in each layer to improve the accuracy of the predictions [19, 20].

(d) Hybrid

The hybrid approach in ANFIS involves using a neural network to adaptively adjust the parameters of a fuzzy logic system. The fuzzy logic system provides a set of rules that map input variables to output variables, and the neural network learns to optimize the parameters of these rules based on a set of training examples. The neural network in ANFIS is typically a type of feedforward network called a backpropagation network, and it is used to adjust the membership functions and rule weights of the fuzzy logic system. The system is also able to learn from data and adapt to new situations, making it a useful tool in a variety of applications, including control systems, pattern recognition, and data analysis [21–24].

2.3 Model Evaluation

The stage consists of evaluation and comparative model algorithm with the optimization model. The result of the prediction process was evaluated using the root mean

Table 1 The result of the root mean square error of the prediction process

No.	Anfis model	RMSE
1	Grid partition backpropagation	0.7059
2	Grid partition hybrid	0.2579
3	Sub-clustering backpropagation	0.7071
4	Sub-clustering hybrid	0.2576

square error (RMSE). RMSE is formulated by:

$$\text{RMSE} = \sqrt{\sum_{i=1}^{n}(t - y)^2} \tag{1}$$

where t is the label value (target prediction) and y is the result of prediction.

3 Result and Reporting Stage

The model prediction implementation used Matlab. The developed ANFIS algorithm model used two generate FIS. There are grid partitions and sub-clustering. The optimization method used backpropagation and hybrid (combination of backpropagation and least square algorithm) for error tolerance 0 with epoch 10. In this research, the load data used the UCI global data. The result of the root mean square error of the prediction process is shown in Table 1.

Table 1 referred to the mean square error for all models, where the lowest error rate is the ANFIS model with a hybrid sub-clustering. It can be concluded that hybrid sub-clustering has better predictive performance or is more accurate in predicting target values than other algorithms. The RMSE is used as an evaluation metric to compare model performance in regression tasks. A lower RMSE value indicates that the model is more accurate at predicting the target value.

4 Conclusion

The cardiovascular diseases prediction model using machine learning has been developed. There are a total of 461 data sets (patient samples). Age, sex, cp, tresbps, chol, fbs, restecg, thalac, exang, oldpeak, and slope are the variables. The algorithm uses a comparative model of grid partitioning and sub-clustering with optimization methods using backpropagation and hybrid (combination of backpropagation and least square algorithm). The results of the analysis of the prediction model for heart disease and stroke resulted in hybrid optimization having the smallest error value, so it can be concluded that sub-clustering with optimization is the best model.

References

1. M. Vaduganathan et al., The global burden of cardiovascular diseases and risk. J. Am. Coll. Cardiol. **80**(25), 2361–2371 (2022). https://doi.org/10.1016/j.jacc.2022.11.005
2. B.C. Lim, et al., Modelling knowledge, health beliefs, and health-promoting behaviours related to cardiovascular disease prevention among Malaysian University students. PLOS ONE **16**(4) (2021). https://doi.org/10.1371/journal.pone.0250627
3. A.-T. Ibtisam, B. Amen, Knowledge and perceived susceptibility of cardiovascular diseases (CVDS) among Saudi female teachers. Int. Arch. Public Health Commun. Med. **4**(2) (2020). https://doi.org/10.23937/2643-4512/1710045
4. Y. Song, et al., Identification of susceptibility loci for cardiovascular disease in adults with hypertension, diabetes and Dyslipidemia & NBSP; [Preprint] (2020). https://doi.org/10.21203/rs.3.rs-75555/v1
5. Y. Tian, Y. Zhang, H. Zhang, Recent advances in stochastic gradient descent in deep learning. Mathematics **11**(3), 682 (2023). https://doi.org/10.3390/math11030682
6. A. Chandrashekhar, U. Desai, P. Abhilash, Cost prediction using gradient descent algorithm. J. Phys. Conf. Series **1706**(1) (2020). https://doi.org/10.1088/1742-6596/1706/1/012038
7. A.M. Husein, A.M. Simarmata, Drug demand prediction model using adaptive neuro fuzzy inference system (ANFIS). SinkrOn. **4**(1), 136 (2019). https://doi.org/10.33395/sinkron.v4i1.10238
8. D. Adyanti, et al., Optimal ANFIS model for forecasting system using different FIS. Proceed. Electr. Eng. Comput. Sci. Inf. **5**(5) (2018). https://doi.org/10.11591/eecsi.v5i5.1617
9. R. Ibrahim, O. Olawale, K. Funmilayo, Diagnosis of hepatitis using adaptive neuro-fuzzy inference system (ANFIS). Int. J. Comput. Appl. **180**(38), 45–53 (2018). https://doi.org/10.5120/ijca2018917020
10. M. Zanganeh, Improvement of the ANFIS-based wave predictor models by the particle swarm optimization. J. Ocean Eng. Sci. **5**(1), 84–99 (2020). https://doi.org/10.1016/j.joes.2019.09.002
11. K. Khosravi, M. Panahi, D. Tien Bui, A comprehensive study of new hybrid models for adaptive neuro-fuzzy inference system (ANFIS) with invasive weed optimization (IWO), differential evolution (DE), firefly (FA), particle swarm optimization (PSO) and bees (BA) algorithms for spatial prediction of groundwater spring potential mapping [Preprint] (2018). https://doi.org/10.5194/hess-2017-707
12. S. Rizvi, et al., A fuzzy inference system (FIS) to evaluate the security readiness of cloud service providers. J. Cloud Comput. **9**(1) (2020). https://doi.org/10.1186/s13677-020-00192-9
13. Y. Huang, et al., Using a machine learning-based risk prediction model to analyze the coronary artery calcification score and predict coronary heart disease and risk assessment. Comput. Biol. Med. **151** (2022). https://doi.org/10.1016/j.compbiomed.2022.106297
14. M. Pishnamazi, et al., Anfis grid partition framework with difference between two sigmoidal membership functions structure for validation of nanofluid flow. Sci. Rep. **10**(1) (2020). https://doi.org/10.1038/s41598-020-72182-5
15. A.A. Ewees, M.A. Elaziz, Improved adaptive neuro-fuzzy inference system using gray wolf optimization: a case study in predicting biochar yield. J. Intell. Syst. **29**(1), 24–940 (2018). https://doi.org/10.1515/jisys-2017-0641
16. B. Selma, S. Chouraqui, H. Abouaïssa, Optimization of ANFIS controllers using improved ant colony to control an UAV trajectory tracking task. SN Appl. Sci. **2**(5) (2020). https://doi.org/10.1007/s42452-020-2236-z
17. C. Federer et al., Improved object recognition using neural networks trained to mimic the brain's statistical properties. Neural Netw. **131**, 103–114 (2020). https://doi.org/10.1016/j.neunet.2020.07.013
18. L.R. Guarneros-Nolasco, et al., Identifying the main risk factors for cardiovascular diseases prediction using machine learning algorithms. Mathematics **9**(20) (2021). https://doi.org/10.3390/math9202537

19. H. Kour, J. Manhas, V. Sharma, Usage and implementation of neuro-fuzzy systems for classification and prediction in the diagnosis of different types of medical disorders: a decade review. Artif. Intell. Rev. **53**(7), 4651–4706 (2020). https://doi.org/10.1007/s10462-020-09804-x
20. M. Nilashi et al., Coronary heart disease diagnosis through self-organizing map and fuzzy support vector machine with incremental updates. Int. J. Fuzzy Syst. **22**(4), 1376–1388 (2020). https://doi.org/10.1007/s40815-020-00828-7
21. H. Moayedi et al., Novel hybrids of adaptive neuro-fuzzy inference system (ANFIS) with several metaheuristic algorithms for spatial susceptibility assessment of seismic-induced landslide. Geomat. Nat. Haz. Risk **10**(1), 1879–1911 (2019). https://doi.org/10.1080/19475705.2019.1650126
22. P. Kora, A. Abraham, K. Meenakshi, Heart disease detection using hybrid of bacterial foraging and particle swarm optimization. Evol. Syst. **11**(1), 15–28 (2019). https://doi.org/10.1007/s12530-019-09312-6
23. M.Z. Abbas, et al., An adaptive-neuro fuzzy inference system based-hybrid technique for performing load disaggregation for residential customers. Sci. Rep. **12**(1) (2022). https://doi.org/10.1038/s41598-022-06381-7
24. D. Tien Bui, et al., New hybrids of ANFIS with several optimization algorithms for flood susceptibility modeling. Water **10**(9) (2018). https://doi.org/10.3390/w10091210

The Smart Concept to Prosper the Community with the Development of Local Wisdom in the *Banjar* Institution

I Dewa Made Adi Baskara Joni, Bazilah A. Talip, Shamsul Anuar Mokhtar, and I Putu Hendika Permana

Abstract The island of Bali is very famous in foreign countries for its tourism. The Balinese economy depends on tourism. However, the global health disaster of the COVID-19 pandemic has devastated Bali's economy. To prevent the pandemic's spread and impact, the government is collaborating with Customary Villages and Customary *Banjar*. The *Banjar* government system is believed to preserve tradition and culture. The existence of *Banjar* can be a driving force for a community-based economy. Developing a smart concept can be a solution to optimize the strength of existing indigenous people with the support of technology. Especially for Bali, it can be developed into a Smart *Banjar*. ICT adoption is a component of the Smart *Banjar*. Qualitative research with interviews was conducted to answer the phenomenon. The findings in this study state that it is necessary to evaluate the level of ICT adoption. Furthermore, it can develop into an appropriate future ICT adoption. The community's proficiency in ICT literacy must be programmed continuously. *Banjar*'s involvement in microenterprise digitization and synergy with community-based microfinance can achieve community welfare.

Keywords ICT adoption · Smart concept · Local wisdom · Prosper the community

I. D. M. A. B. Joni (✉) · B. A. Talip · S. A. Mokhtar
Universiti Kuala Lumpur Malaysian Institute of Information Technology, 50250 Kuala Lumpur, Malaysia
e-mail: dewadi@unmas.ac.id

B. A. Talip
e-mail: bazilah@unikl.edu.my

S. A. Mokhtar
e-mail: shamsulanuar@unikl.edu.my

I. D. M. A. B. Joni
Information Technology Department, Universitas Mahasaraswati Denpasar, Denpasar 80233, Indonesia

I. P. H. Permana
Digital Business Department, Institut Bisnis dan Teknologi Indonesia, Denpasar 80225, Indonesia
e-mail: hendika@instiki.ac.id

© The Author(s), under exclusive license to Springer Nature Switzerland AG 2024
A. Ismail et al. (eds.), *Tech Horizons*,
SpringerBriefs in Applied Sciences and Technology,
https://doi.org/10.1007/978-3-031-63326-3_12

1 Introduction

Today, the development of technology is unstoppable. All aspects of life have developed with the support of technology, especially information communication technology (ICT). The Corona virus disease (COVID-19) pandemic has accelerated the use of ICT. Starting from the central and local governments, the business sector, several organizations, and schools utilize ICT when activity restrictions occur. The pandemic has increased both the speed and the population's use of Internet channels [1]. However, the pandemic has had a negative impact on Bali's economy, which has tourism as its primary sector.

Bali is a unique island. Bali is famous for its world-class tourism. Foreign tourist arrivals before the COVID-19 pandemic continued to increase every year. Foreign tourist visits 2015 were recorded to be 4,927,937 and 6,275,210 in 2019 [2]. The data does not include visits by domestic tourists, which have a more significant number. Nevertheless, poverty remains a problem behind the development of Bali's tourism. In Bali, 3.79% of residents are considered poor as of March 2019 [3]. In Indonesia, especially in Bali, the government is collaborating with Customary Villages and Customary *Banjar* to prevent and handle the spread and impact of a pandemic [4].

Bali has two kinds of villages: *Desa Adat* (Customary Village) and *Desa Dinas* (Official Village). Since the time of the monarchy, the Balinese have followed a village administration history [5]. The *Banjar*, under the direction of *Kelian Banjar*, is the Village's lowest unit of government [6, 7]. The *Banjar* has existed since the tenth to eleventh centuries [8, 9]. The *Banjar* government system is believed to preserve the traditions and culture of the Balinese people. *Banjar* is an institution that has proven to survive for thousands of years. Based on the strength of the *Banjar*, it is a potential that can be developed from time to time. The Customary Village and Customary *Banjar* have special autonomy rights and can manage their households. Customary *Banjar* can manage the community's traditions, religious, and socio-economic activities. The institution manages various aspects of people's lives, from tradition to socio-economic activities. At each *Banjar*, there is a *Bale Banjar* facility (a place where the community carries out activities, including customary activities). *Banjar,* located in urban and tourist areas, expands the *Bale Banjar*'s function for economic activity [7, 10]. Several *Bale Banjars* have an automated teller machine (ATM), mini market, and rented out for retail stores. Despite these transformations, with *Bale Banjar* serving as a public area, Banjar continues to serve as a stronghold defending tradition and socio-cultural society. Consequently, the outcomes can benefit the community [8].

Social inequality caused by poverty can be overcome by strengthening the socio-economic community [11]. The culture of *gotong royong* and local genius *asah, asih,* and *asuh* can become the basic philosophy to address the changing times' challenges. The existence of *Banjar* as a local wisdom can be a driving force for a community-based economy. Developing a smart concept can be a solution to optimize the strength of existing indigenous peoples with the support of appropriate technology. The smart concept can be developed in the lowest structure bottom-up. Especially for Bali, with its traditions and local wisdom, it can be developed into a

Smart *Banjar*. Each *Banjar,* microenterprises, and microfinance institutions collaborate to realize the Smart *Banjar* objective [12]. The smart concept developed will synergize microenterprises and microfinance to prosper the community. This synergy will support local wisdom, like the *Banjar* Institution.

2 Methodology

Qualitative researchers choose from among the possibilities, such as narrative, phenomenology, ethnography, case study, and grounded theory [13]. This research involves three elements or entities. These elements or entities are *Banjar*, microenterprise, and microfinance. The three elements or entities will synergize and form a Smart *Banjar* concept. Smart *Banjar* will be realized with an ICT adoption. Smart *Banjar* is an undeveloped idea that must be tailored to the community's conditions, desires, and requirements. This can only be done if it involves participant action in the research. Researchers believe that viewpoints, culture, and historical background greatly influence understanding the research object. This research relies on the participants' views about the situation being studied [13].

2.1 Data Collection

Good research is research that can be proven empirically. Such proof requires appropriate research techniques and procedures. This research requires a collection of facts in the field. It is what is called data. Participatory observation techniques are frequently combined with in-depth interviews in qualitative research. Throughout the duration of the investigation, the researcher conducted interviews with the subjects [14]. The following will explain the data collection techniques in this research. Based on the methodology chosen and the research strategy that has been determined, the data collection technique that has been done is interviews.

2.2 Data Analysis

Qualitative data analysis is inductive. The data are used to do the analysis, which is then turned into a hypothesis. In qualitative research, data analysis is done before going into the field, in the field, and after the fieldwork is done [14]. Processing the data will use the assistance of qualitative computer software such as Atlas.ti 9. The qualitative data analysis process that has been carried out refers to [13] as follows: (1) Organize and prepare the data for analysis; (2) Read or look at all the data; (3) Start coding all of the data; (4) Generate a description and themes; (5) Representing the description and themes.

3   Results and Discussion

Finding and defining problems, figuring out formulations and goals, and planning a research method were the first steps in research. The research is conducted in a structured manner, fulfills scientific principles, and is proven empirically. Data collection and research were conducted in Kukuh Village, Kerambitan District, Tabanan Regency, Bali Province, Indonesia. Data were obtained by direct interview techniques with 35–45 min durations. Based on this, there are results and findings in this study. These findings have novelty, contribute to the body of knowledge, and benefit stakeholders. Several findings were obtained from the data that was collected and analyzed. These findings constitute initial information and knowledge related to the smart concept in developing local wisdom, in this case, the *Banjar* Institution in Bali.

3.1   ICT Adoption

In the analysis related to ICT adoption, there are two main findings. The findings are associated with current ICT adoption and future ICT adoption. Several things are part of the current ICT adoption: a payment system, e-complaint, administration system, chatting application, ICT for information dissemination, and service acceleration. Then there are millennials microenterprise ICT adoption, a phenomenon where the younger generation can adopt ICT to run their business. However, contradictions in several places, types of services, and community groups still need ICT adoption. In future ICT adoption, several parts are expected, such as microfinance admin feature, microfinance private feature, microfinance public feature, order system, and online marketplace. Of course, it is hoped that the technological devices will have user-friendly functions. For the utilization of future ICT adoption to run well, it is necessary to increase people competencies such as capacity building, coaching, and upgrading skills. For more details, see Fig. 1.

3.2   Smart Concept

The smart concept is intended to solve urban problems and can be developed for rural areas [15]. The Garuda Smart City Framework 2 (GSCF 2) tools have been developed in Indonesia. GSCF 2 has indicators including digital government as part of the measurement, apart from sustainable indicators such as economy, social, and environment. In addition, GSCF 2 also has enabler indicators such as technology or infrastructure, people, and governance [16]. The Government of Indonesia has declared 350 villages as smart village locus for 2021 as part of the Village Development Strategy [17]. The findings in this study stated that several things are part of the smart village, such as smart farming, digitize data of MSME, and smart people.

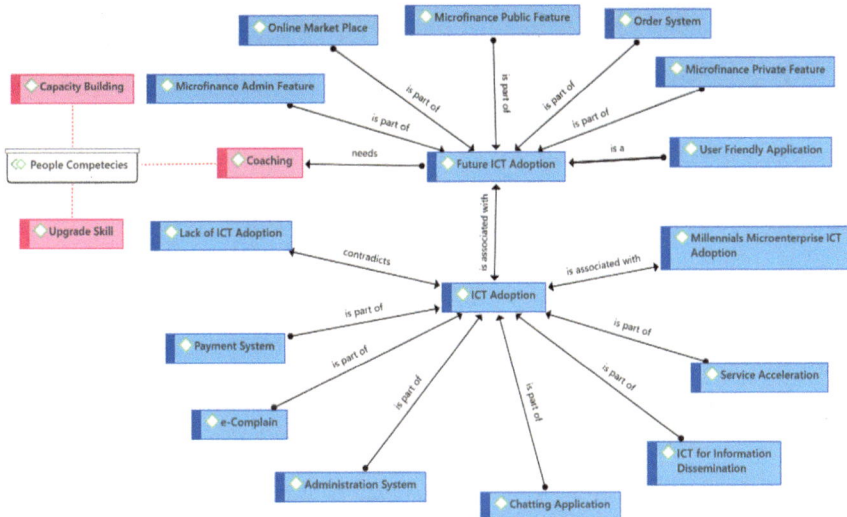

Fig. 1 ICT adoption

Farmer empowerment and the empowerment of agricultural products are part of smart farming, which is the head of village policy. *Banjar* is the data source for digitizing data from micro, small, and medium enterprises (MSMEs). Digital community and digital literacy are the cause of smart people. There are centralized funds in the village, village development programs, and village centralized programs in community development. Village-owned enterprises (BUMDesa) synergy with *Banjar* is related to *Banjar* as a community entity that is part of the village's centralized program. Smart village of Indonesia is a factor that causes prosper the community. This is related to people's welfare with the impact on village development caused by BUMDesa synergy with microenterprise and community-based microfinance, the optimal community-based economy. For more details, see Fig. 2.

3.3   Smart Concept with ICT Adoption

This section is a result of the smart concept development with ICT adoption. Most of the findings have been described in Sects. 3.1 and 3.2 above. Several additional findings are in this section, as in the administration system, which is part of the current ICT adoption. The administration system is associated with *Banjar* as a community entity. This is because the *Banjar* is an institution with a role and responsibility for administering the legal needs of its people, especially on official *Banjar*. The official village and *Banjar* in Bali are extensions of the central government in administering their citizens [5]. Then on community-based microfinance which is stated as

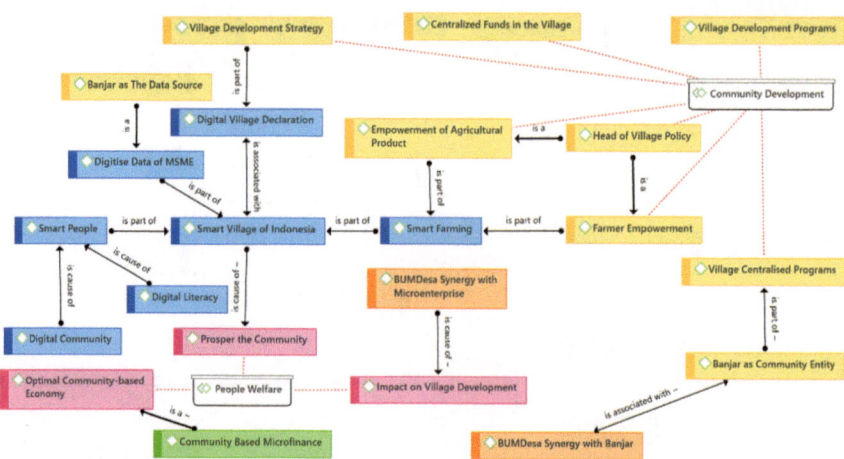

Fig. 2 Smart concept

optimal community-based economy. It has supporting sections such as the microfinance admin feature, microfinance private feature, and microfinance public feature, which are part of future ICT adoption. For more details, see Fig. 3.

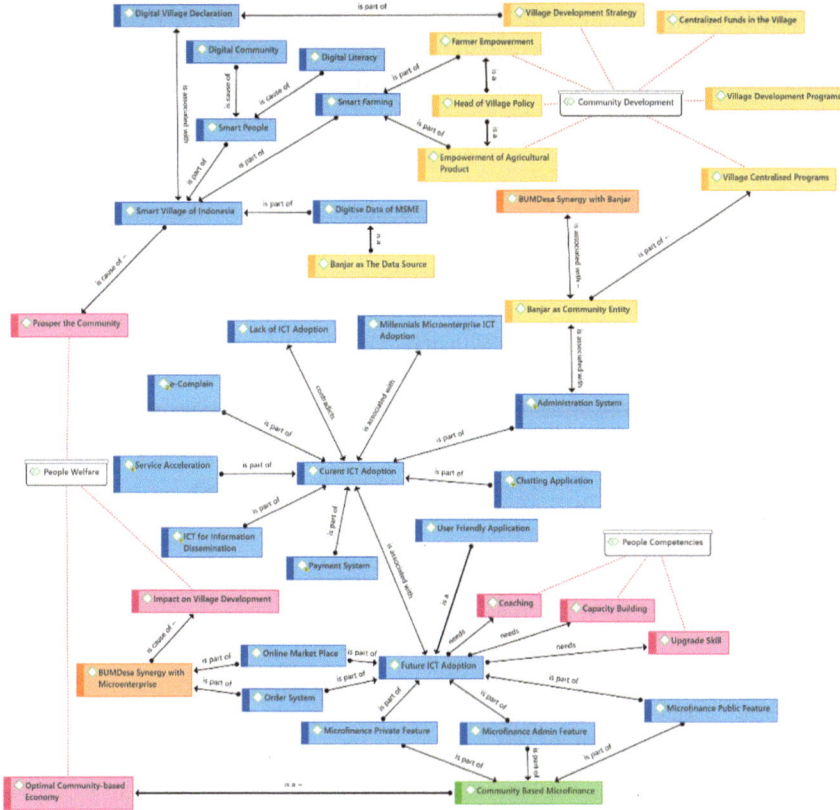

Fig. 3 Smart concept with ICT adoption

4 Conclusion

Research has been conducted with a structured methodology to answer the problems and achieve the objectives. Analysis and empirical evidence are obtained based on the data collected from several roles, such as head of village, head of *Banjar*, microfinance leaders, and microenterprise leaders. Smart concepts have generally been developed to answer urban (Smart City) and rural (Rural Smartness/Smart Village) problems. The application of the smart concept in the development of local wisdom has begun. The research seeks to develop a framework that enables the synergy of microenterprise, microfinance, and *Banjar* by leveraging ICT to reduce external parties' dependence on the community's fundamental requirements [12]. In adopting an ICT solution, it is necessary to consider the current level of ICT adoption. The service must be available starting from the administration system to improve services and provide e-Complaint. The initial foundation is the availability of millennials microenterprise ICT adoption. However, to be developed further, it is necessary to assess the level of ICT adoption in each region. The required future ICT adoption is developed after obtaining vital information related to current ICT adoption. This starts from features in microfinance to handling orders on the online marketplace. Of course, all features developed must be user-friendly. Next, people competencies such as capacity building, coaching, and upgrading skills must be programmed to ensure sustainability. Based on the results and research findings, it is necessary to digitize the MSME data and *Banjar* as the data sources in the development of smart village. Because government programs are centralized in the village, it is necessary to involve *Banjar* as a community entity that can be synergized with BUMDesa. For the welfare of society, it is necessary to involve community components and groups that can become an optimal community-based economy. It can be done by synergizing with community-based microfinance. Finally, the smart concept to prosper the community with the development of local wisdom can be realized.

Acknowledgements We want to thank Universiti Kuala Lumpur, Universitas Mahasaraswati, and Institut Bisnis dan Teknologi Indonesia for facilitating the resources for research. We also thank Kukuh Village, Kerambitan, Tabanan-Bali, for participating in this research.

References

1. L.E. Relihan, M.M. Ward Jr., C.W. Wheat, D. Farrell, The early impact of COVID-19 on local commerce: changes in spend across neighborhoods and online. SSRN Electron. J. (2020). https://doi.org/10.2139/ssrn.3647298
2. Central Bureau of Statistics Bali Provinces: Pertumbuhan Ekonomi Bali Tahun 2019 (2020)
3. Central Bureau of Statistics Bali Provinces: Profil Kemiskinan di Bali Maret 2019. Denpasar (2019)
4. I.G. Januariawan, Pencegahan Covid 19 Berbasis Desa Adat di Desa Adat Tanggahan Peken Perspektif Hukum Adat, in *COVID-19: Perspektif Hukum dan Sosial Kemasyarakatan* ed. by I.B.S. Saitya, I.M.P. Subawa (Yayasan Kita Menulis, Medan, 2020), p. 63
5. I.G. Parimartha, *Silang Pandang Desa Adat dan Desa Dinas di Bali* (Udayana University Press, Denpasar, 2013)
6. I.W. Suasnawa, I.M.S.A. Jaya, I.B.K. Sugirianta, PKM Pengelolaan Keuangan Banjar Adat Di Desa Sangeh. Provinsi Bali. J. Bhakti Persada. **4**, 65–75 (2018)
7. N.P. Suwardani, W. Paramartha, I.G.A. Suasthi, Bale Banjar and its implications on the existence of Bali socio-cultural communities, in *Proceeding Book International Seminar on Tolerance and Pluralism in Southeast Asia The Faculty Of Religious and Cultural Science*. The Faculty of Religious and Cultural Science Universitas Hindu Indonesia, Denpasar (2018), pp. 83–90
8. N.P.A. Sawitri, W.H. Nugrahandika, Tipologi Perkembangan Pemanfaatan Lahan Bale Banjar dan Faktor-Faktor yang Memengaruhinya Studi Kasus Kota Denpasar, Provinsi Bali, in *Seminar Nasional Space #3* (Universitas Hindu Indonesia, Denpasar, 2017), pp. 352–376
9. A.A.N.A. Sanjaya, I.N.H. Juliarthana, Pemanfaatan Bale Banjar Sebagai Ruang Kreativitas Anak Muda Di Kota Denpasar. J. Sp. **1**, 26–32 (2019)
10. C. Gantini, H. Hanan, The impact of tourism industry on the sustainability of traditional Bale Banjar in Denpasar, in *Proceedings of the 6th International Conference of Arte-Polis* (2017)
11. M. Syawie, Kemiskinan dan Kesenjangan Sosial. J. Inf. **16**, 213–219 (2011)
12. I.D.M.A.B. Joni, S.A. Mokhtar, B.A. Talip, Conceptual framework of smart Banjar: a synergize of the micro-enterprise, microfinance, and The Banjar Institution. Adv. Econ. Bus. Manag. Res. **202**, 7–13 (2021). https://doi.org/10.2991/aebmr.k.211226.002

13. J.W. Creswell, J.D. Creswell, *Research Design Qualitative, Quantitative, and Mixed Methods Approaches* (SAGE Publications, Inc., All, 2018)
14. Sugiyono: Metode Penelitian Kualitatif. Alfabeta, Bandung (2021)
15. R.N. Andari, S. Ella, Developing a smart rural model for rural area development in Indonesia. J. Borneo Adm. **15**, 41–58 (2019)
16. S.H. Supangkat, A.A. Arman, R.A. Nugraha, Y.A. Fatimah, The implementation of Garuda smart city framework for smart city readiness mapping in Indonesia. J. Asia-Pac. Stud. **32**, 169–176 (2018)
17. Ministry of Villages Development of Disadvantaged Regions and Transmigration, R. of I.: 350 Desa Lokus Desa Cerdas Tahun 2021. https://kemendesa.go.id/berita/assets/files/pengum uman_desa_cerdas_1_page_(1).pdf

Laser-Based Security Monitoring Alarm Triggered System in Industrial Application Using IoT

Nurhusna Muhamad Nazari, Noor Hidayah Mohd Yunus, Hafiz Basarudin, and Norliana Yusof

Abstract There is a potential risk of theft and trespassing occurring on industrial premises because various business operations are handled. Several methods are adopted to deal with this issue, such as using the services of security guards and installing security alarms for security purposes. However, it involves a less cost-effective electronic system. Therefore, this security monitoring system that uses lasers was proposed for the purpose of dealing with this issue. Laser light is used to cover large areas because lasers can travel long distances without scattering effects, have sufficient energy to trigger the security system in a small zone and the existence of laser light is not clearly noticeable compared to the cable connection concept. Thus, intruders cannot detect that there is a security alarm installed in the area. This project aims to design and develop a security alarm laser-triggered system using an Arduino UNO microcontroller and IoT technology interface with a camera that could monitor the movement through camera footage on mobile apps. The main components used are an Arduino to execute commands and a web camera that displays a video when the intruder crosses the laser. The proposed project is beneficial for recent security issues regarding high-rated intruders on industrial premises.

N. M. Nazari · N. H. M. Yunus (✉)
Advanced Telecommunication Technology, Communication Technology Section, Universiti Kuala Lumpur British Malaysian Institute, Batu 8, Jalan Sungai Pusu, 53100 Gombak, Selangor, Malaysia
e-mail: noorhidayahm@unikl.edu.my

N. M. Nazari
e-mail: nurhusna.nazari15@s.unikl.edu.my

H. Basarudin
Lee Kong Chian Faculty of Engineering and Science, Universiti Tunku Abdul Rahman, Kampar, Selangor, Malaysia
e-mail: hafizba@utar.edu.my

N. Yusof
Faculty of Innovative Design and Technology, Universiti Sultan Zainal Abidin, 21300 Kuala Terengganu, Terengganu, Malaysia
e-mail: norliana@unisza.edu.my

Keywords Wireless camera and router · Laser · Light sensor intensity · CCTV ·
Arduino UNO

1 Introduction

Security systems are vital for everyone to secure homes, offices and warehouse areas
for peaceful living, help protect against criminals, and prevent incidents such as unau-
thorized access and intrusion. In the late 1980s, the earliest security system namely
closed-circuit television (CCTV) was introduced, it was a very valued commodity
for everyone at that time and more and more people installed such systems to stay
protected and safe [1, 2]. However, there were drawbacks that the security system
was physically visible that could alert the intruder and the system was also without an
alarm monitoring system to alert the owner of the subsequent action. Owing to these
drawbacks, it opens more opportunities for intruders to commit attempted crimes,
leading to an unprecedented growth in the crime rate.

To overcome this problem, the development of a laser-based security system and
the intruder is unaware that a security system is installed in the entry positions like
doors or windows is introduced. In this development, sensor-based using Internet of
thing (IoT) is implemented in applications such as laser light sensing assistance for
security purposes, wireless monitoring, and notification alert through mobile apps.
In the technological era of industrial revolution 4.0 (IR 4.0), sensor-based, and IoT
applications aim to assist and facilitate all aspects of human life [3–6].

Laser-based security systems are implemented due to applications which are
capable of covering larger areas such as industrial warehouse areas. The laser waves
can go through a long distance with controllable light rays, travel in phase and have
less scattering effect [7]. The laser light rays emit narrow rays of penetrating electro-
magnetic radiation and can go in concentrated energy in between 650 nm wavelength
range in a straightway and are almost invisible [8, 9]. This is why laser rays are very
narrow and can be focused into a tiny spot or become invisible. Laser ray offers to
direct intense light in remarkable ways that can only be seen at the source and point
of incidence with coherent length in a few kilometers. These laser characteristics
inspired the development of laser-based security systems.

In this paper, the proposed laser-based security system uses the IoT technology
and adopts an Arduino UNO microcontroller with a piezo alarm buzzer connected
and a camera capturing images. When any person or intruder crossover the laser light,
the alarm buzzer starts ringing and the authorized person can access monitoring alerts
with the captured image of the intruder through mobile apps connected with Wi-Fi.
The proposed laser-based security system shown in Fig. 1 is further explained in the
following section.

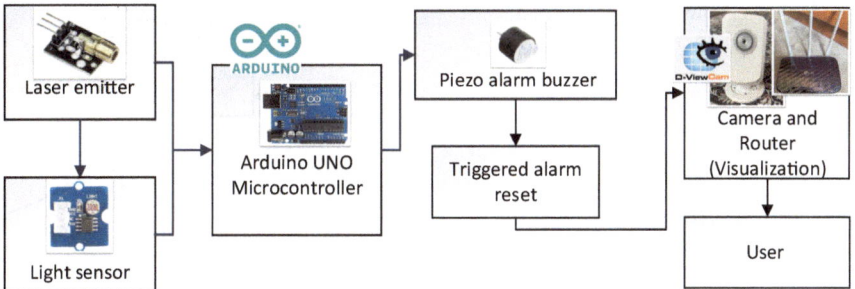

Fig. 1 Block diagram of the system

2 Methodology

The block diagram of the laser-based security system is shown in Fig. 2. On the front end of the system, the proposed laser-triggered alarm security system consists of a light emitter and a light sensor as a sensing unit when the circuit is triggered. An Arduino UNO is used to make the coding and control all the movement as well as ray detection. The alarm buzzer gives an alert when the intruder has passed in the monitored entrance and camera for image visualization. On the back end of the monitoring system side, a smartphone with an Internet connection has functioned to view the captured image of the intruder.

The operational flow of the system is shown in Fig. 3. The process starts with checking every corner side that needs to place a mirror, then initializing the laser emitter at the transmitter side. The emitter radiates laser rays to the receiver side, consisting of a light depending resistor (LDR) sensor connected with the driver circuit. The LDR sensor senses the laser ray continuously while the microcontroller unit begins to process the signal. When the person or intruder crossover the laser ray, thus the LDR sensor is triggered due to discontinuity of the signal from the laser ray. Then, the camera is activated for visualization records of the intruder imaging to the microcontroller unit. Due to discontinuity laser ray, the alarm circuit is auto-activated and keeps on active mode until the user manually pushes the reset button. Finally, notification alarms and video camera recordings are displayed through a mobile application connected to the system.

3 Results and Discussion

The most important part of the system in industrial premises application is the intensity of the single laser light ray scattered over a long distance. The laser ray is an almost invisible boundary of a sensitive area and is only visible at the radiation and incident points. Thus, a mirror can function to make the laser ray visible by reflection concept, the placement of the mirrors at every corner as illustrated in Fig. 4. The

Fig. 2 Project flow chart

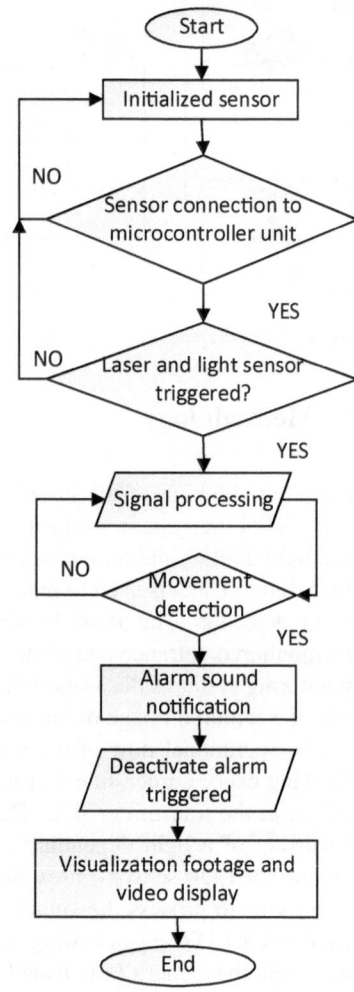

advantage of the method is that power consumption is low and high visibility of the laser ray. An illustration of a laser ray being reflected by a mirror is shown in Fig. 5.

The detected light intensity parameter in terms of brightness and darkness from the video camera footage is shown in Fig. 6. When the surrounding conditions are dark, the light intensity is an average of 47 W m^{-2}, while the average value is 373.1 W m^{-2} when the surrounding conditions are bright. Intensity is the power delivered by the light of the surface area. The intensity of the light, I is based on the speed of light in the vacuum, $c = 3.00 \times 10^8$ m/s, the directed light at 100% reflective surface and the light pressure, p in unit N/m^2. The formula to calculate the intensity of light, I is as follows [10, 11]:

From the basic pressure formula,

Fig. 3 Laser-based security
monitoring system network

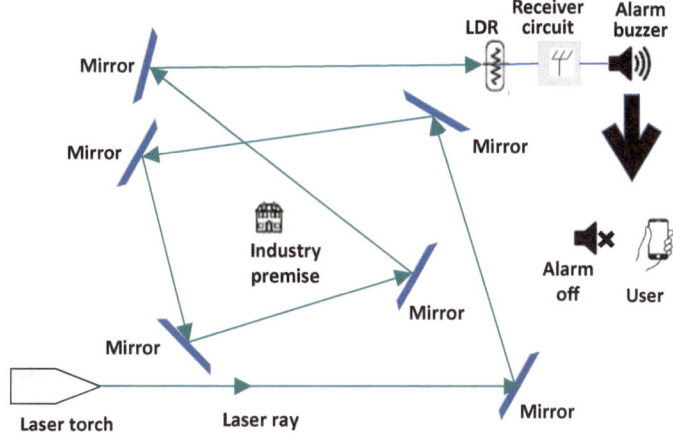

Fig. 4 Connection of laser security system for industry premise

$$\text{Pressure, } P = \frac{\text{Force, } F}{\text{Area, } A} \tag{1}$$

$$\text{Force, } F = \frac{\Delta P}{\Delta t} \tag{2}$$

The formula for calculating the radiation pressure for a perfectly reflective surface,

$$\text{Radiation pressure, } P = \frac{2I}{c} \tag{3}$$

Thus,

Fig. 5 The laser ray
reflected onto the mirrors

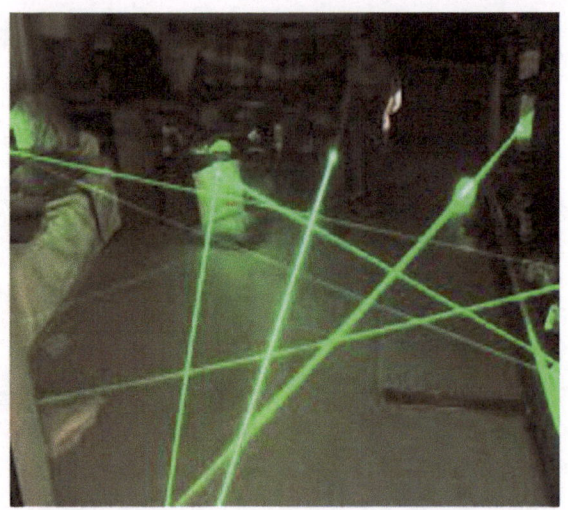

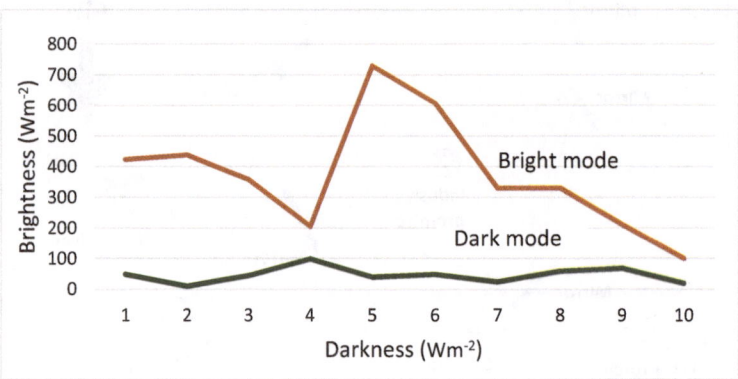

Fig. 6 Light intensity of bright and dark modes

$$\text{Intensity of light, } I = \frac{cP}{2} \tag{4}$$

4 Conclusion

This paper proposed the development of a laser-triggered alarm security system
in industrial applications using Arduino and IoT technology interfaces. Nowadays,
security systems are vital to prevent theft and intruders and also for security asset
purposes. Experimental implementation and testing with the laser light and a light
sensor have been conducted several times to demonstrate the mediator between the

sensors and the MCU to activate the alarm buzzer. The project has achieved a functional security alarm triggered by laser light and a trespasser monitoring system. However, the limitation of this system is the requirement of a stable Wi-Fi network to transmit recorded video to the users from the camera in real time.

Future work should be directed to research wireless data transmission network infrastructure that can adapt to the long-range transmission distance of Wi-Fi signals interconnected by a router in accessing multiple devices. In the future, LoRa RFID-based transmission technology and IoT platforms could be well implemented [12–14], allowing them to operate over longer distances of more than 2 km at lower costs and with less power consumption.

Acknowledgements The authors wish to thank Universiti Kuala Lumpur British Malaysian Institute and Center for Research and Innovation, CoRI of Universiti Kuala Lumpur, Malaysia for the support given to the success of this project.

References

1. C.E. O'Neill, The artist as surveillant: the use of surveillance technology in contemporary art. Doctoral dissertation, Sotheby's Institute of Art-New York (2022)
2. H. Lu, Z.D. Xu, T. Iseley, et al., *Pipeline Inspection and Health Monitoring Technology: The Key to Integrity Management* (Springer Nature, 2023)
3. Z. Khaslan, N.H.M. Yunus, M.S.M. Nadzir et al., IoT-based indoor air quality monitoring system using SAMD21 ARM cortex processor. Adv. Mater. Eng. Technol. **162**, 245–253 (2022)
4. M.A. Kamarudin, N.H.M. Yunus, M.R.A. Razak et al., Development of Blynk IoT platform weather information monitoring system. Adv. Mater. Eng. Technol. **162**, 295–305 (2022)
5. M.I.A. Suhaidi, N.H.M. Yunus, Development of Blynk IoT-based air quality monitoring system. J. Eng. Technol. **9**, 63–68 (2021)
6. N. Yusof, S.M. Norzeli, S.N. Yusof et al., Modeling and simulation of planar micro-coils for invasive pressure sensing. Adv. Mater. Sci. Tech. Led Women **165**, 165–172 (2023)
7. F. Wang, F. Ren, Z. Ma et al., In vivo non-invasive confocal fluorescence imaging beyond 1,700 nm using superconducting nanowire single-photon detectors. Nat. Nanotechnol. **17**, 653–660 (2022)
8. D.L. Andrews, *Lasers in Chemistry* (Springer Science & Business Media, 2012)
9. X. Luo, D. Tsai, M. Gu et al., Extraordinary optical fields in nanostructures: from sub-diffraction-limited optics to sensing and energy conversion. Chem. Soc. Rev. **48**(8), 2458–2494 (2019)
10. J. Agil, R. Battesti, C. Rizzo, On the speed of light in a vacuum in the presence of a magnetic field. Eur. Phys. J H **48**(1), 2 (2023)
11. J. Zhou, P. Li, J. Wang, Effects of light intensity and temperature on the photosynthesis characteristics and yield of lettuce. Horticulturae **8**(2), 178 (2022)
12. N. Azmi, L.M. Kamarudin, A. Zakaria, et al., Radio frequency identification (RFID) range test for animal activity monitoring, in *IEEE International Conference on Sensors and Nanotechnology* (2019), pp. 1–4
13. L.M. Kamarudin, A. Zakaria, M.N. Rahman, et al., Monitoring feeding and resting pattern of goats in dairy farm using long-range RFID-based system, in *Proceedings of the 7th International Conference on Communications and Broadband Networking* (2019), pp 41–45
14. M.T. Bakar, A.A. Jamal, Determination of suitable resource discovery tool and methodology for high-volume internet of things (IoT). J. Phys. Conf. Series. **1874**(1), 012046 (2021)

Short-Term Photovoltaic (PV) Energy Prediction Using Machine Learning Approach

Norzanah Md Said, Raja Fazliza Raja Suleiman, Noor Hasyimah Abu Rahim, and Mohd Juhari Mat Basri

Abstract The efficient prediction of short-term photovoltaic (PV) energy output is a pressing challenge in the renewable energy sector. Accurate PV energy forecasts are pivotal for optimizing grid integration, minimizing energy wastage, and reducing operational costs in solar power plants. This study addresses these challenges by leveraging machine learning (ML) techniques and comparing the performance of three ML models, namely linear regression, random forest, and gradient boosting, with the objective of identifying the most effective model for short-term PV energy prediction during the given timeframe. The study utilizes 5 MWp PV power plant data collected in Melaka, Malaysia, over a daily period from 1st September 2013 until 31st January 2014 providing a robust dataset for training and testing the models. The primary evaluation metrics used in this analysis are the root mean squared error (RMSE), R-squared (R^2) score, and the mean absolute percentage error (MAPE). The findings reveal that the gradient boosting (GB) model outperforms both linear regression (LR) and random forest (RF) in terms of predictive accuracy; RMSE (1380.13), R-squared (R^2) (0.8), and MAPE (4.3%). This suggests that GB is the most suitable ML model for accurate short-term PV energy prediction in the context of Melaka's PV power plant data.

N. M. Said (✉) · R. F. R. Suleiman · N. H. A. Rahim
Industrial Automation Section, Universiti Kuala Lumpur Malaysia France Institute, Bangi, Selangor, Malaysia
e-mail: norzanah@unikl.edu.my

R. F. R. Suleiman
e-mail: fazliza@unikl.edu.my

N. H. A. Rahim
e-mail: noorhasyimah@unikl.edu.my

M. J. M. Basri
Water and Energy Section, Universiti Kuala Lumpur Malaysia France Institute, Bangi, Selangor, Malaysia
e-mail: mohdjuhari@unikl.edu.my

N. M. Said · R. F. R. Suleiman · N. H. A. Rahim
UniKL Robotics and Industrial Automation Center, Universiti Kuala Lumpur Malaysia France Institute, Bangi, Selangor, Malaysia

© The Author(s), under exclusive license to Springer Nature Switzerland AG 2024   111
A. Ismail et al. (eds.), *Tech Horizons*,
SpringerBriefs in Applied Sciences and Technology,
https://doi.org/10.1007/978-3-031-63326-3_14

Keywords PV energy prediction · Machine learning models · Gradient boosting · Renewable energy

1 Introduction

Recently, digital technologies and automation have transformed the energy landscape, particularly photovoltaic (PV) energy production and monitoring [1]. Smart technologies and data-driven strategies have been incorporated into energy-generating and distribution operations as a result of industry 4.0 [2, 3]. This paradigm shift is crucial as the world uses more renewable energy like sunlight to produce electricity [4]. Harnessing PV energy efficiently demands innovative technologies and reliable resources, making it essential to make the transition to sustainable energy [5].

One of the key challenges that has surfaced with the growing adoption of solar PV systems is the need for accurate short-term PV power forecasts. Such forecasts are indispensable for the successful integration of solar energy into existing power grids, the optimization of energy storage solutions, and the reduction of operational costs [6, 7]. However, short-term PV power prediction poses a significant hurdle in the realm of renewable energy. Numerous methods and models have been explored in scholarly literature [8, 9], often combining numerical weather prediction (NWP) models with statistical techniques [10]. These predictions typically rely on factors such as PV irradiation, temperature, and historical PV power data, but statistical postprocessing models have struggled to capture localized fine-grained variations.

In response to these challenges, the field of renewable energy research has increasingly turned to machine learning (ML) approaches. Studies have demonstrated that ML algorithms [11], including support vector regression (SVR) [12] and gradient boosting [13], hold promise in capturing the complex and nonlinear relationships between weather conditions and PV power generation. The focus of this research is on short-term PV energy forecasts in Melaka, Malaysia. Using PV power plant data, this study analyzes and forecasts PV energy output. To estimate daily PV energy output, this study examines three ML models, namely linear regression (LR), random forest (RF), and gradient boosting (GB). Comparative analysis is used to find the most efficient model for short-term PV energy forecasting. Data collection, preprocessing, ML models, assessment metrics, and results analysis and discussions are covered in subsequent sections.

2 Methodology

2.1 Dataset of PV Power Generation

A 5 MWp PV power plant data in Melaka, Malaysia, provided PV output energy and PV insolation data for this study. Valuing filtering uses meteorological parameters such as PV panel temperature and ambient temperature. The data is recorded daily from 1 September 2013 to 31 January 2014.

An extensive data preparation was done before analysis, to ensure data consistency and correctness, data preprocessing included deleting duplicate entries, missing values, and outliers using imputation. To train the learning model, the data is split into two subsets, the training set and the testing set, with a distribution ratio of 70% and 30%, respectively.

2.2 Machine Learning as a Learning Model

2.2.1 Linear Regression

Linear regression is a fundamental and widely used regression model in machine learning and statistics for establishing a linear relationship between the independent variables (input features) and the dependent variable (PV power production) [14]. In this study, linear regression was applied as a baseline model for short-term PV energy prediction.

2.2.2 Random Forest

Random forest is an ensemble learning method that combines multiple decision trees to make predictions. The algorithm constructs a multitude of decision trees at training time and outputs the mean or average prediction of the individual trees for regression tasks [15]. In this study, random forest was chosen for its ability to capture intricate relationships between meteorological variables and PV energy production.

2.2.3 Gradient Boosting

Gradient boosting (GB) is one of the ensemble techniques that builds predictive models sequentially by combining multiple weak learners, typically decision trees, to create a strong predictive model [16]. The algorithm trains the model sequentially, and each new model tries to correct the previous model, resulting in a strong learner that can handle complex and nonlinear relationships. The algorithm computes the gradient of the loss function with respect to the predictions of the current ensemble and then trains a new weak model to minimize this gradient. In this study, GB was chosen for its high predictive accuracy and adaptability to complex data patterns.

2.3 Prediction on PV Output Energy Performance Indices

To assess the performance of the ML models, this study employs the following evaluation metrics known as root mean squared error (RMSE), R-squared (R^2) score, and the mean absolute percentage error (MAPE) [17]. These metrics are widely recognized as indicators of predictive accuracy and model performance.

$$RMSE = \sqrt{\frac{1}{n} * \sum (y_i - \hat{y}_i)^2}, \tag{1}$$

$$R^2 = 1 - \left(\frac{\sum (\hat{y}_i - \overline{y})^2}{\sum (y_i - \overline{y})^2} \right), \tag{2}$$

$$MAPE = \frac{1}{n} * \sum \left(\frac{y_i - \hat{y}_i}{y_i} \right) * 100, \tag{3}$$

where n is the number of data points (samples) in the PV panel plant dataset, y_i is the actual (observed) values, *PV output energy*, $\hat{y}_i$ is the predicted values generated by ML models for the testing dataset, $\overline{y}$ is the mean (average) of the actual values in the PV panel plant dataset. The selected prediction model meets the criteria for prediction accuracy. Lower RMSE, a higher R^2 score, and the smallest MAPE values indicate a better fit to the data and more accurate predictions. Results of the tested model performances using Eqs. (1)–(3) are given in Table 1.

Table 1 Comparison of prediction models performances for short-term PV output energy

Evaluation parameters	Supervised learning methods		
	Gradient boosting	Random forest	Linear regression
RMSE	1380.13	1495.376	1862.134
R^2	0.80	0.76	0.63
MAPE (%)	4.3	4.9	6.5

3   Results and Discussion

3.1   Descriptive Statistical Analysis of Meteorological Parameters and PV Output Energy

Figure 1 shows the statistical summary for PV insolation levels, the daily average in degrees Celsius (°C) for PV panel, ambient temperature, and PV output energy. According to the box and whisker plots, most of the data points fell within the interquartile range, indicating stable meteorological parameters throughout the investigation. Note that no variable had outliers. This shows the dependability of PV insolation, PV panel temperature, ambient temperature, and PV output energy data. The dataset's integrity, modelling accuracy, and results reliability are improved by eliminating extreme values. Thus, solar energy generation and use forecasts will be more accurate.

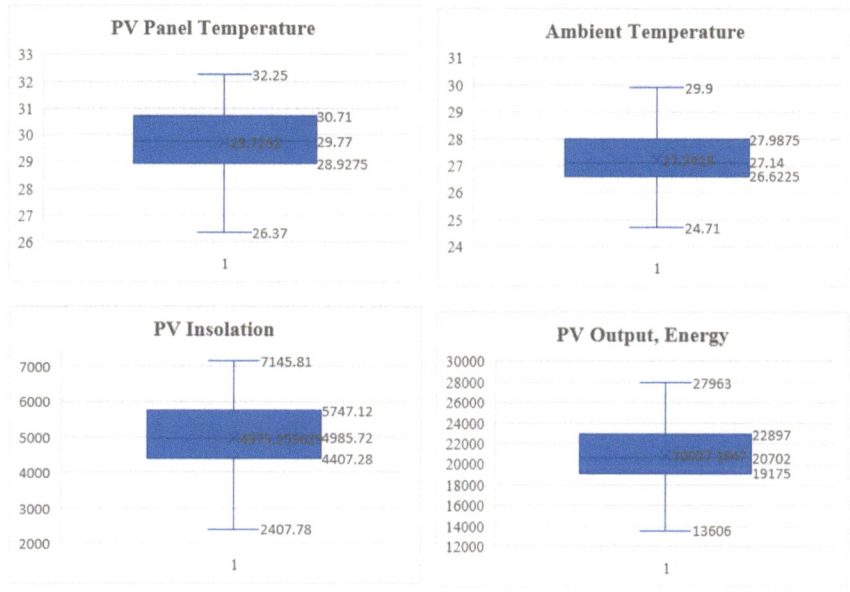

Fig. 1 Descriptive statistical analysis of meteorological parameters and PV output energy via box and whisker plots

3.2 PV Output Energy Prediction

3.2.1 Model Evaluation Metrics

Three ML models are assessed using RMSE, R^2, and MAPE metrics to assess data fit and prediction accuracy. Table 1 shows that the gradient boosting (GB) model has performance metrics of 1380 RMSE, 0.80 R^2, and 4.3% MAPE. The random forest (RF) model has an RMSE of 1495.37, R^2 of 0.76, and MAPE of 4.9%. The linear regression (LR) model has an RMSE of 1862.14, R^2 of 0.63, and MAPE of 6.5%.

According to the results, the gradient boosting model outperforms the random forest and linear regression models in all three performance criteria. It has the lowest RMSE (1380), best R^2 (0.80), and smallest MAPE (4.3%), indicating higher accuracy and data fit.

3.2.2 Actual vs. Predicted PV Output Energy

Figure 2 shows the line plots of the model's predictions versus PV output energy levels. In the line plot for GB, the predicted PV output energy nearly matches the actual numbers, showing the model's ability to capture data patterns and variances. The closeness of the predicted line to the actual line demonstrates better accuracy in short-term PV energy prediction using GB. The RF line plot also matches expected and actual PV output energy levels. This suggests that the RF model makes reliable predictions. Although there may be tiny differences, the forecasts approximate the values. However, the line plot for LR fits the data well, with predicted values often following the trend of actual values. The variability and data fit of LR are slightly worse than those of GB and RF.

Overall, these line plots show each model's predicted performance. The most accurate short-term PV output energy prediction model is GB, followed by RF. LR provides useful insights into variable relationships, but it seems less accurate at predicting this difficulty.

4 Conclusion

Accurate short-term PV energy prediction is vital for effective grid integration and efficient utilization of solar energy resources. In this study, three machine learning models, namely linear regression, random forest, and gradient boosting were employed to forecast PV output energy for a 5 MWp PV power plant installed in Melaka, Malaysia.

In this study, the analysis revealed that GB outperforms the other models, achieving the lowest RMSE, highest R^2 score, and the smallest MAPE. Consequently, this study demonstrates that a GB model can be a valuable tool for predicting PV

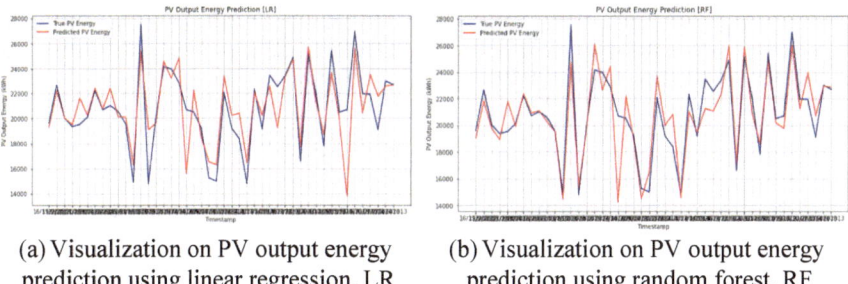

(a) Visualization on PV output energy prediction using linear regression, LR

(b) Visualization on PV output energy prediction using random forest, RF

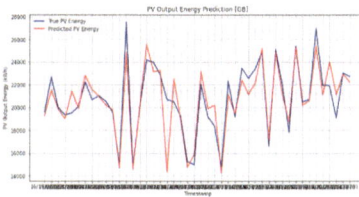

(c) Visualization on PV output energy prediction using gradient boosting, GB

Fig. 2 Actual versus predicted PV output energy based on **a** LR, **b** RF, and **c** GB model

output energy based on meteorological parameters in the context of Melaka's PV power plant data. The results indicate its potential for practical applications in the solar energy sector, but ongoing research and model refinement are necessary to address complex relationships and achieve even more accurate predictions. Future research could explore more advanced machine learning techniques, such as neural networks or deep learning, to potentially improve the accuracy of PV output energy predictions.

References

1. H.L. Kangas, K. Ollikka, J. Ahola, Y. Kim, Digitalisation in wind and solar power technologies. Renew. Sustain. Energy Rev. **150**, 111356 (2021). https://doi.org/10.1016/j.rser.2021.111356
2. P. Buła, T. Schroeder, M. Ziółko, Renewable energy through industry 4.0 on the example of photovoltaic development in selected European countries, in *The Future of Management*, ed. by B. Nogalski, P. Buła (Jagiellonian University Press, 2022), pp. 92–107
3. V. Pandey, A. Sircar, N. Bist, K. Solanki, K. Yadav, Accelerating the renewable energy sector through industry 4.0: optimization opportunities in the digital revolution. Int. J. Innov. Stud. **7**(2), 171–188 (2023). https://doi.org/10.1016/j.ijis.2023.03.003
4. S. Nižetić, N. Djilali, A. Papadopoulos, J.J.P.C. Rodrigues, Smart technologies for promotion of energy efficiency, utilization of sustainable resources and waste management. J. Clean. Prod. **231**, 565–591 (2019). https://doi.org/10.1016/j.jclepro.2019.04.397
5. K. Hercegová, T. Baranovskaya, N. Efanova, Smart technologies for energy consumption management. SHS Web Conf. **128**, 02005 (2021). https://doi.org/10.1051/shsconf/202112 802005

6. D. Niu, K. Wang, L. Sun, J. Wu, X. Xu, Short-term photovoltaic power generation forecasting based on random forest feature selection and CEEMD: a case study. Appl. Soft Comput. **93**, 106389 (2020). https://doi.org/10.1016/j.asoc.2020.106389

7. NREL, *Forecasting Wind and Solar Generation: Improving System Operations* (2015)

8. M.B.M. Juhari et al., Statistical moments approach in grid-connected photovoltaic system performance evaluation. Appl. Mech. Mater. **785**, 616–620 (2015)

9. B. Espinar et al., Photovoltaic forecasting: a state of the art to cite this version, in *5th European PV-Hybrid Mini-Grid Conference*, Apr 2010, Tarragona, Spain (2010), pp. 250–255, [Online]. Available: https://hal-mines-paristech.archives-ouvertes.fr/hal-00771465

10. C. Voyant, M. Muselli, C. Paoli, M.-L. Nivet, Numerical weather prediction (NWP) and hybrid ARMA/ANN model to predict global radiation. Energy **39**(1), 341–355 (2012). https://doi.org/10.1016/j.energy.2012.01.006

11. D. Markovics, M.J. Mayer, Comparison of machine learning methods for photovoltaic power forecasting based on numerical weather prediction. Renew. Sustain. Energy Rev. **161**, 112364 (2022). https://doi.org/10.1016/j.rser.2022.112364

12. M. Alrashidi, S. Rahman, Short-term photovoltaic power production forecasting based on novel hybrid data-driven models. J. Big Data **10**(1), 26 (2023). https://doi.org/10.1186/s40537-023-00706-7

13. J. Wang, P. Li, R. Ran, Y. Che, Y. Zhou, A short-term photovoltaic power prediction model based on the gradient boost decision tree. Appl. Sci. **8**(5) (2018). https://doi.org/10.3390/app8050689

14. S. Ng, et al., An insight of linear regression analysis. Sci. Res. J. **15**, 1. https://doi.org/10.24191/srj.v15i2.5477

15. M. Schonlau, R.Y. Zou, The random forest algorithm for statistical learning. Stata J. **20**(1), 3–29 (2020). https://doi.org/10.1177/1536867X20909688

16. J.H. Friedman, Greedy function approximation: a gradient boosting machine. Ann. Stat. **29**(5), 1189–1232 (2001). [Online]. Available: http://www.jstor.org/stable/2699986

17. K. Ramasubramanian, J. Moolayil, Applied supervised learning with R: use machine learning libraries of R to build models that solve business problems and predict future trends. Packt Publishing Ltd. (2019)

Insider Threat Prediction Techniques: A Systematic Review Paper

Nur Fahimah Mohd Nassir, Ummul Fahri Abdul Rauf, Zuraini Zainol, and Kamaruddin Abdul Ghani

Abstract The aim of this study is to increase comprehension and provide a systematic review of insider threat prediction techniques explored by previous researchers. The advantages and disadvantages of machine learning techniques, statistical techniques, hybrid techniques, and knowledge-based techniques are highlighted. In addition, prospective work obstacles and suggestions have been discussed. This study examined insider threat prediction trends in scholarly articles published between 2007 and 2022. Researchers, practitioners, and policymakers who are interested in predicting insider threats should find this study beneficial.

Keywords Insider threat · Mitigation approach · Prediction technique · PRISMA methodology

N. F. M. Nassir
Department of Defence Science, National Defence University of Malaysia, Kem Sungai Besi, 57000 Sungai Besi, Malaysia
e-mail: 3231786@alfateh.upnm.edu.my

U. F. A. Rauf (✉)
Department of Mathematics, National Defence University of Malaysia, Kem Sungai Besi, 57000 Sungai Besi, Malaysia
e-mail: ummul@upnm.edu.my

Z. Zainol
Department of Computer Science, National Defence University of Malaysia, Kem Sungai Besi, 57000 Sungai Besi, Malaysia
e-mail: zuraini@upnm.edu.my

K. A. Ghani
Department of Electrical and Electronics Engineering, National Defence University of Malaysia, Kem Sungai Besi, 57000 Sungai Besi, Malaysia
e-mail: kamaruddin@upnm.edu.my

© The Author(s), under exclusive license to Springer Nature Switzerland AG 2024
A. Ismail et al. (eds.), *Tech Horizons*,
SpringerBriefs in Applied Sciences and Technology,
https://doi.org/10.1007/978-3-031-63326-3_15

1   Introduction

The rapid growth of information technology and the digital domain has improved business, banking, healthcare, transportation, and government [1]. Along with the benefits, digital data and computer systems have increased cybersecurity risks [2], including malware, phishing, man-in-the-middle, passwords, social engineering, advanced persistent threats, distributed denial-of-service (DDoS) attacks, and SQL injection attacks [3]. While organisations invest heavily in protecting their valuable information from external threats, there is still an increasing concern about insider threats, which can be more challenging to identify and prevent and may cause severe consequences [4]. According to [5], insider-caused incidents have increased, with 67% of organisations having 21–40 every year. In contrast, just 53% and 60% of organisations reported 21–40 events in 2018 and 2020.

Insider threats refer to "the potential for an individual who has or had authorised access to an organisation's assets to use their access, either maliciously or unintentionally, to act in a way that could negatively affect the organisation", and this individual can be a current or former employee, contractor, or business partner [6]. Based on [7], these threats can cause significant financial and reputational damage. Insider attacks can also compromise sensitive data's confidentiality, integrity, and availability, leading to data leaks, IP theft, and industrial espionage [8].

It is worth noting that researchers have studied insider threat mitigation strategies, including prediction, detection, and prevention. Figure 1 shows the Scopus-analysed publication frequency of insider threat mitigation measures from 2003 to May 2023. This timeline was chosen because insider threat prediction research is not always preferred by academics and is underutilised to mitigate insider threats. Even though prediction improves decisions, reduces uncertainty, and improves outcomes, many academics confuse it with sophisticated methodologies or narrowly forecast the future, which may lead to unfavourable outcomes [9].

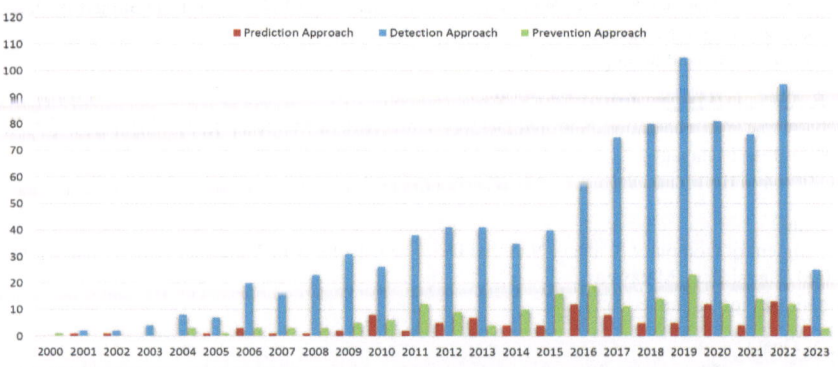

Fig. 1 Investigated shapes of projectiles (geometry and dimensions)

Due to the above scenarios, this study seeks to improve understanding and give a systematic review of insider threat prediction techniques. The PRISMA methodology was utilised to assure the review process's quality and reliability. First, relevant publications were identified through a search engine-based literature search. The data extraction process was then used to screen and exclude the article based on the study criteria. Above all, this study strengthened the evidence basis, informed decision-making, guided future research, and advanced knowledge in numerous disciplines.

2 Techniques for Insider Threat Prediction

2.1 Machine Learning Technique

The authors [10] proposed a novel predictive model called DANTE that employs a long short-term memory (LSTM) neural network to predict insider threats from system logs. The model is trained and evaluated using user login, file access records, and system events. By using log sequence information, DANTE can identify abnormal user behaviour more accurately than traditional methods. On the other hand, [11] highlights the significance of user behaviour while using a file system to identify insider threats. The author described how machine learning algorithms, such as decision trees, random forests, and support vector machines can analyse server message block (SMB) network behaviour, which involves SMB servers, clients, and a network switch. Whereas [12] proposed simulating intellectual property theft with a game-based method to examine insider threat behaviour and personality. The authors explore how personality traits, facial expressions, and language can predict insider threats. Additionally, in [13], the authors proposed a static or mobile device network activity monitoring system. The authors also discuss threat level, which measures insider threat severity and risk associated with an individual or behaviour.

2.2 Statistical Technique

In [14], a statistical model that considers technical, organisational, and human factor perspectives using a Bayesian network for predicting insider threats has been developed. Despite using a multi-perspective approach, Nebrase [15] developed a comprehensive insider threat risk assessment in order to assist organisations in identifying their weaknesses in handling insider threats. This study utilised real data from one educational organisation to create a dynamic Bayesian network model. Similar to [15, 16] outlined malicious insiders' motivation and psychology. The authors have created a unique structural equation model and used psychological elements to predict the degree of interest in a potentially malicious insider with a Bayesian network model. Furthermore, in [17], a system dynamics model and a Bayesian belief network were integrated. Personality, attitude, and counterproductive behaviour are used in the system dynamics model to simulate insider attacks. Then, the Bayesian belief network is built to predict attack risk by combining key variables from the system dynamics model and learning the conditional probabilities from simulated scenarios. The study examined risk indicators for a normal employee, an openly dissatisfied malicious insider, and a malicious insider who hides bad behaviour. Other than that, [4] analysed human behaviours through system logins and recognised unusual behaviour. This study was predicted using the regression method. In addition, the authors provided a baseline of normal user behaviours to compare to unusual behaviours to predict and identify potential insider threats.

2.3   Hybrid Technique

In the context of forensic investigation, Wei et al. [18] introduced an unsupervised anomaly detection strategy that used statistical and machine learning methods on system logs. The approach uses real-world datasets to analyse user behaviours, resource accesses, and system events. Since the method does not depend on labelled datasets, it is flexible and adaptable to changing threat scenarios. On the other hand, in [19], the authors propose combining the insider trust profile matrix (TPM) with document sensitivity values to generate a risk matrix to predict insider threats. The risk value, prediction rate, and risky path were calculated and analysed using an insider threat prediction framework. The authors used the e-plantation system (ePS) as a case study to show that the suggested method can alarm breach occurrences and predict insider threats.

2.4   Knowledge-Based Technique

Domain specific language (DSL) and insider threat prediction and specification language (ITPSL) have been proposed in [20] to profile typical misuse incidents and express threat detection and prediction systematically. The prediction technique used in this study is called "evaluated potential threat (EPT)". EPT employs a set of rules and expert knowledge to evaluate the potential threat of an event based on identified conditions. Despite, in [21], the authors addressed the issue of insider threats in relational database systems and classified dependencies and constraints that insiders can use to extract confidential data. New tools introduced in the study, such as the algorithm for constructing insiders' knowledge graphs (CDG), dependency matrix, and threat prediction graph might assist organisations in preventing data leakage.

Table 1 Criteria for inclusion and exclusion in the primary studies

Inclusion criteria	Exclusion criteria
The paper should focus on the insider threat prediction	Other insider threat mitigation approaches, such as detection, prevention, and protection are the focus of the paper
The paper must highlight prediction techniques	Article published earlier than 2003
The paper should focus on the prediction technique's effectiveness, limitations, challenges, and suggestions for future work	The paper does not adequately discuss the findings

3 Methodology

The articles pertaining to insider threat prediction have been identified from impact journals, conference proceedings, and book sections through the use of a search engine-based literature search. Search terms were "cybersecurity", "insider threat", and "prediction". To choose relevant publications, this study used PRISMA, which comprises four phases: identification, screening, eligibility, and inclusion. The platform's database search and other sources found 161 studies. This number was reduced to 101 after removing duplicate research. Then, 44 publications were examined after rigorous inclusion and exclusion criteria evaluation, as indicated in Table 1. After data extraction, 27 articles were eliminated for accuracy and relevancy. To achieve the review's objectives, the remaining 17 publications were analysed.

4 Results and Discussion

This section discusses the outcomes of applying the previously mentioned methodology, classifying the pertinent studies by year, and analysing the number of relevant publications for each reviewed year and their origin. Figure 2 displays the annual number of relevant studies published between 2007 and 2022. This 15-year period was selected to match relevant sources. The graph shows a consistent interest in insider threat prediction techniques over the past few years, with a publication peak in 2022. However, the number of articles published in different years varies, indicating shifts in research emphasis and attention over time. It is essential to note that the number of published articles in the early 2000s was relatively low. Yet, as the field of insider threat prediction gained recognition and significance, research activity began to increase gradually.

Machine learning techniques are typically the most accurate, but training and evaluating machine learning models requires a large dataset of historical data. In previous studies, behavioural analysis, game-based methods, and surveillance were also employed. Behavioural analysis methods are less precise than machine learning techniques, but they can be used to identify threats in real time and do not require

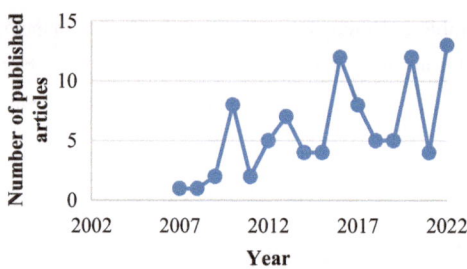

Fig. 2 Frequency of papers published per year

a large historical dataset. Game-based methods are a comparatively novel approach to predicting insider threats, but they have the potential to be more accurate than traditional approaches. In contrast, methods of surveillance can be highly effective at identifying suspicious behaviour, but they can also be intrusive and raise privacy concerns. The limitations of this technique are that data can be difficult to obtain, as many organisations are reluctant to share their security logs with researchers. Additionally, labelling can be time-consuming and error-prone, while scalability is crucial for deployment in large organisations [12].

According to previous studies, the Bayesian network was utilised by the majority of the prior researchers who used statistical techniques to predict insider threats. However, the most appropriate approach for predicting insider threats will depend on the organisation's particular requirements and model. Bayesian network models may be most effective for organisations that possess large amounts of historical data and the resources to deploy them. Compared to organisations that need to identify threats in real time or do not have recourse to a significant amount of historical data, they may benefit more from using regression or prediction intervals. On the other hand, organisations that seek to comprehend the path to an insider attack may be better served by using system dynamics models. The availability of data, the complexity of the models, and the cost of deployment are all limitations involved in this technique.

Several insider threat prediction techniques have relied only on technical data, such as system logs and network traffic. However, it has been demonstrated that these techniques are limited in their ability to predict insider threats [10]. Therefore, this technique has been proposed as a hybrid technique that combines statistical analysis, machine learning, and behavioural analysis to improve prediction accuracy. In addition, these studies provide insight into the various strategies that can be used to predict insider threats and identify the key factors that must be considered when developing a model for insider threat prediction.

Knowledge-based techniques employ information regarding insider behaviour and access privileges to detect anomalies and predict insider threats. ITPSL is effective at predicting insider threats at the file, network, process execution, and hardware device levels, but it requires comprehensive knowledge about insider behaviour and dependencies, which can be challenging to obtain. On the other hand, the CDG and TPG are effective at detecting and preventing threats, and they can be used to improve security and mitigate risks associated with insider attacks. Yet it has only been evaluated on limited datasets, and it is unclear whether it is appropriate for larger datasets, and its performance in a real-world setting is unclear.

5   Conclusion and Future Study

This study focused on insider threat prediction techniques, which must be improved for organisations to mitigate risks from intentional and unintentional insiders. Without comprehensive and accurate datasets, which are often sensitive and confidential, insider threat analysis is challenging. Obtaining tagged data for analysis is also difficult. To overcome this difficulty, future studies should provide larger and more diverse datasets that include multiple perspectives, such as behavioural, technical, and organisational. This study also found that many risk analysis methods rely on a single perspective and insufficient data. It should be noted that insider threat prediction frequently involves the analysis of sensitive personal data, which may violate privacy rights. Hence, future research should consider and explore ways to develop privacy-preserving techniques and frameworks that can achieve a balance between the need for accurate prediction and the protection of individual privacy. In addition, this study suggests that future studies should consider hybrid techniques where the researchers can develop comprehensive and efficient models using a combination of machine learning, statistical analysis, and knowledge-based techniques. Numerous methods have shown potential in controlled environments, but their real-world usefulness is in doubt. Due to that, field experiments and case studies to assess these strategies' operational efficacy should be focused on in future studies. In conclusion, this study should shed light on this topic and encourage more effective prediction method research. This review might assist organisations and researchers in creating more accurate and robust insider threat prediction tools, improving organisational security by filling this knowledge gap.

References

1. I. Singh, N. Kaur, Contribution of information technology in growth of Indian economy. Int. J. Res. Granthaalayah 5, 1–9 (2017). https://doi.org/10.29121/granthaalayah.v5.i6.2017.1986
2. S. Quach, P. Thaichon, K.D. Martin, S. Weaven, R.W. Palmatier, Digital technologies: tensions in privacy and data. J. Acad. Mark. Sci. 50, 1299–1323 (2022). https://doi.org/10.1007/s11747-022-00845-y
3. O. Aslan, S.S. Aktuğ, M. Ozkan-Okay, A.A. Yilmaz, E. Akin, A comprehensive review of cyber security vulnerabilities, threats, attacks, and solutions. Electronics (Basel) 12, 1333 (2023). https://doi.org/10.3390/electronics12061333
4. J.U. Mills, S.M.F. Stuban, J. Dever, Predict insider threats using human behaviors. IEEE Eng. Manage. Rev. 45, 39–48 (2017). https://doi.org/10.1109/EMR.2017.2667218
5. 2022 Cost of Insider Threats Global Report (2022)
6. Common Sense Guide to Mitigating Insider Threats 7th Edition (2022)

7. Insider Threat Mitigation Guide (2020)
8. J. Eggenschwiler, I. Agrafiotis, J.R. Nurse, Insider threat response and recovery strategies in financial services firms. Comput. Fraud Secur. **2016**, 12–19 (2016). https://doi.org/10.1016/S1361-3723(16)30091-4
9. M.D. Verhagen, A pragmatist's guide to using prediction in the social sciences. Socius **8** (2022). https://doi.org/10.1177/23780231221081702
10. Q. Ma, N. Rastogi, DANTE: predicting insider threat using LSTM on system logs, in *Proceedings—2020 IEEE 19th International Conference on Trust, Security and Privacy in Computing and Communications, TrustCom 2020*, Institute of Electrical and Electronics Engineers Inc. (2020), pp. 1151–1156
11. N.K. Niemann, R.G. Blockmon, Naval Postgraduate School Monterey, California Thesis using Machine Learning to Predict the Insider Threat in a Network Environment (2021)
12. S. Basu, Y.H. Victoria Chua, M. Wah Lee, W.G. Lim, T. Maszczyk, Z. Guo, J. Dauwels, Towards a data-driven behavioral approach to prediction of insider-threat, in *Proceedings—2018 IEEE International Conference on Big Data, Big Data 2018*. Institute of Electrical and Electronics Engineers Inc. (2019), pp. 4994–5001
13. K. Bhavsar, B.H. Trivedi, An insider cyber threat prediction mechanism based on behavioral analysis, in *Advances in Intelligent Systems and Computing* (Springer, 2016), pp. 345–353
14. N. Elmrabit, S.H. Yang, L. Yang, H. Zhou, Insider threat risk prediction based on Bayesian network. Adv. Intell. Syst. Comput. **96** (2020). https://doi.org/10.1016/j.cose.2020.101908
15. E. Nebrase, *A Multiple Perspective Approach for Insider Threat Risk Prediction in Cyber-Security* (2018)
16. E.T. Axelrad, P.J. Sticha, O. Brdiczka, J. Shen, A Bayesian network model for predicting insider threats, in *Proceedings—IEEE CS Security and Privacy Workshops*, SPW 2013 (2013), pp. 82–89
17. P.J. Sticha, E.T. Axelrad, Using dynamic models to support inferences of insider threat risk. Comput. Math. Organ. Theory **22**, 350–381 (2016). https://doi.org/10.1007/s10588-016-9209-1
18. Y. Wei, K.P. Chow, S.M. Yiu, Insider threat prediction based on unsupervised anomaly detection scheme for proactive forensic investigation. Forensic Sci. Int. Digit. Invest. **38** (2021). https://doi.org/10.1016/j.fsidi.2021.301126
19. I. Ismail, R. Hassan, M. Razib Othman, A. Syifaa' Ahmad, N. Elya Tawfiq, Insider risk profile matrix to quantify risk value of insider threat prediction framework. J. Theor. Appl. Inf. Technol. **15**, 19 (2017)
20. G. Magklaras, S. Furnell, The insider threat prediction and specification language, in *Ninth International Network Conference—INC2012* (2012)
21. Q. Yaseen, B. Panda, *Predicting and Preventing Insider Threat in Relational Database Systems*